CONSTRUCTION AND DESIGN MANUAL

建造设计手册

建筑竞赛点评与图解

COMPETITION PANELS AND DIAGRAMS

［德］本杰明·胡斯巴赫
［德］克里斯蒂安·雷姆豪斯 编
［德］克里斯丁·艾希曼

胡一可 张天翔 译

江苏凤凰科学技术出版社

图书在版编目（CIP）数据

建造设计手册：建筑竞赛点评与图解／（德）胡斯巴赫，（德）雷姆豪斯，（德）艾希曼编；胡一可，张天翔译．－－南京：江苏凤凰科学技术出版社，2016.4
 ISBN 978-7-5537-6213-5

Ⅰ．①建⋯ Ⅱ．①胡⋯ ②雷⋯ ③艾⋯ ④胡⋯ ⑤张⋯ Ⅲ．①建筑设计－欧洲－图解 Ⅳ．①TU206

中国版本图书馆CIP数据核字（2016）第053747号

建造设计手册：建筑竞赛点评与图解

编 者	［德］本杰明·胡斯巴赫
	［德］克里斯蒂安·雷姆豪斯
	［德］克里斯丁·艾希曼
译 者	胡一可 张天翔
项目策划	凤凰空间／陈 景
责任编辑	刘屹立
特约编辑	林 溪
出版发行	凤凰出版传媒股份有限公司
	江苏凤凰科学技术出版社
出版社地址	南京市湖南路1号A楼，邮编：210009
出版社网址	http://www.pspress.cn
总 经 销	天津凤凰空间文化传媒有限公司
总经销网址	http://www.ifengspace.cn
经 销	全国新华书店
印 刷	深圳市建融印刷包装有限公司
开 本	1 020 mm×1 440 mm 1／16
印 张	24
字 数	269 000
版 次	2016年4月第1版
印 次	2016年4月第1次印刷
标准书号	ISBN 978-7-5537-6213-5
定 价	388.00元（精）

图书如有印装质量问题，可随时向销售部调换（电话：022-87893668）。

建造设计手册
建筑竞赛点评与图解

[德]本杰明·胡斯巴赫
[德]克里斯蒂安·雷姆豪斯 编
[德]克里斯丁·艾希曼

胡一可 张天翔 译

目 录

6 前 言

12 竞赛随笔
14 建筑的瑰宝——"映像历史":建筑竞赛及其表达
22 参赛成本
32 竞赛绘图表达
54 个人竞赛方案演示
58 竞赛成果制作、提交及其他细节

60 竞赛项目
62 旧城复兴 —— Lalla Yeddouna 广场
82 马尼托巴大学校园及周边地区
100 越南 - 德国大学校园
118 河内科技大学校园
136 蒂森克虏伯房地产有限公司办公楼
150 50 Hertz 公司总部大楼
158 BMW FIZ Future
184 威斯巴登法学院(隶属于欧洲商业学校)
194 神经退行性疾病中心
204 尤里西生物研究所生物研究园
214 埃森大学附属医院儿科门诊及核医学研究大楼
228 联邦警察总部大楼
238 席勒公园房地产开发
246 花园城区法尔肯堡居住区
256 下萨克森州议会大厦
264 卡尔广场邻近地区
284 拉尔日护中心
290 诺伊尔市场(新罗斯托克北部市场)
298 住宅 F
306 "建筑的艺术"——勃兰登堡州议会大厦
310 吕讷堡大学校园景观设计
320 新贝斯科艺术档案馆
332 莫斯科理工博物馆
348 Mystetskyj 兵工厂文化区
362 洲际酒店、滑冰俱乐部、音乐厅
374 其他竞赛项目

376 附 录
378 索 引
384 致 谢

前　言

竞赛表达

设计与建造的过程中存在着各个层面、多种形式的沟通渠道，它们是设计与建造过程的一大特色。信息交流呈现多元化的形式，例如，与客户会面、电视电话会议、虚拟数据库、电子邮件、与当地议员和市民团体进行晚间会议以及前往政府部门或执行董事会收集资料并与建造商交谈等。项目从开始直到交付使用（有时甚至更久）一直伴随着此类事务。除语言沟通外，设计理念与想法的交流还可以采用多种方式，例如，模型、草图、平面图、技术图纸等。项目的第一阶段——"准备阶段"在某种程度上可以视为信息交流过程的重中之重。因此，项目的成功依赖于有效的沟通。

然而，在项目进程中还存在着不同寻常的沟通阶段（或称为"缺乏沟通的阶段"）。在这个阶段，参赛者（建筑师、规划师）以个人意愿构思设计，与竞赛主办方（竞赛之后将变成业主）并没有直接的沟通。虽然这种"背对背"状态在建筑竞赛中是必要的，但其会让双方感到紧张和沟通不畅。对于竞赛主办方而言，"失语"或交流的阻断是难以接受的，就如同在未知的黑暗隧道的尽头才能看到最终的判决。然而，参赛者对于"背对背"状态的态度则是矛盾的，他们还是愿意相信，只有在独立的环境氛围中，个人的创作理念和设计作品才能得到充分表达。

大多数竞赛组织机构都理解并且支持"匿名体系"所具有的公正性，因此上述竞赛进程中的信息交流方式一直延续着。

竞赛任务书通常是一本图文并茂的小册子，其中列出了主办方想要告知参赛者的所有细节：总体意图和实施过程的具体条款。在设计阶段，交流仅仅发生于讨论会和相应的问答环节。讨论内容不得涉及最终提交作品的图纸和模型。于是，图纸和模型的重要性远远超过建筑师的演讲口才和个人魅力。

在这种情况下，方案设计必须十分出色，以至于仅仅几平方米的图纸就能传达参赛者希望评委理解的所有设计意图，并

吸引他们为其投票。方案设计的展示成为业主与建筑师建立互信关系的基础。方案设计不仅根据项目的现实条件进行空间的规划设计，同时还是一种互助共享式的设计解读与操作。类似规模的基础建设同样适用于投资类项目，比如具有多种使用功能的公共设施；很多个人建筑和企业建筑的项目也都期待空间的规划设计最终能够带来经济效益或实现经济上的飞跃。然而，其所面临的挑战是：越走向跨界且受跨文化影响的项目，越需要与之对应的全面、及时的信息交流。

左页：德国拉尔日护中心建筑竞赛评审会初审现场
上图：德国慕尼黑 BMW FIZ Future 建筑竞赛第二次市民大会中，
　　　一位技术专家正在陈述设计意图
下图：德国柏林 50 Hertz 公司总部大楼建筑竞赛评审现场

在这个阶段，方案设计的表达工具包括图形、文字、统计数据以及带比例的模型。方案表达的核心部分是各层平面，即建筑设计意向图解和特殊形态的空间平面展示。

前文所描述的"特殊的沟通方式"催生了一种高端的图纸表达文化。竞赛要求参赛者在寻找针对方案设计的最佳解决策略的同时，运用平面图解、文字图表等进行综合、多元、精彩的表达。虽然评委不应该被图纸表达所蒙蔽，但图纸的易读性和吸引力的确在评审过程中起到至关重要的作用。展示并诠释图纸表达的杰出案例是本书的重点。

上图：德国多特蒙德小教堂，炭笔和粉笔描图纸（设计：汉斯·夏隆）
右页：俄罗斯加里宁格勒州因斯特堡市政大厅，炭笔描图纸（设计：汉斯·夏隆）

设计与建造指南

设计与建造指南分为两部分：第一部分重点阐述了建筑竞赛的一般性问题，即通过纸质或其他媒介的表达，以及在某些情况下的现场演示。此外，一些表达技巧和基本框架也被列入其中。第二部分列举了近7年来笔者所在的工作室（[phase eins]. 工作室）组织举办的建筑竞赛中25份设计作品的演示文件。竞赛主办方对建筑品质的重视与把控以及参赛者无与伦比的方案展示共同成就了这些作品。在对建筑竞赛的总体介绍中，Eva-Maria Barkhofen教授阐述并展现了图纸表达的历史脉络。为了使讨论的话题更加广泛，在技术层面，笔者以"竞赛作品如何表达以及为何如此表达？如何审视并评价这些表达"为话题，邀请了很多业内同行、业主和记者陈述观点与看法。这些问题及其解答引发了业内同行所期望的讨论：选择正确、恰当的表达方式作为应对问题的策略手段，以及在全球范围内对于建筑竞赛系统现状的看法。

在建筑竞赛中，图纸表达已在某种程度上成为竞赛表达的发展潮流和趋势，平面图尤其如此。与本书所列举的表达方式相比，如果有的话，透视图、配色、尺寸、图面布局和字体的重要性也只能被部分评价。二者的区别在20世纪70年代、20年代甚至更早的19世纪建筑竞赛系统中体现得更加明显。绘画等表现技法的变革所产生的影响就是一个很好的例证：原先使用铅笔和油墨，后来发展为钢笔，再后来发展为模板、计算机甚至三维打印技术。与技术发展同步的是视觉世界的影像和人们的观看习惯的变化，这些都影响着普通人和专家群体的评价标准。由此，建筑竞赛的状况也在发生改变，以不断适应人们对于精致图像的需求和认同。人们不但接受如此劳民伤财地制作渲染图，而且还把它作为评价标准。显然，这会引发一系列争议，就如同对于渲染图缺乏真实感的讨论。

下图：德国慕尼黑奥林匹克体育公园，竞赛方案概念草图，1967（设计：Günter Behnisch 及合伙人事务所，绘制：Carlo Weber）
右图：德国柏林亚历山大广场，城市设计竞赛方案，1993（设计：Libeskind 工作室）

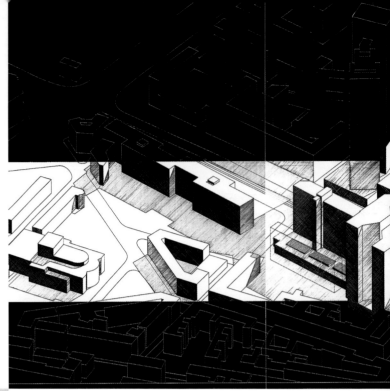

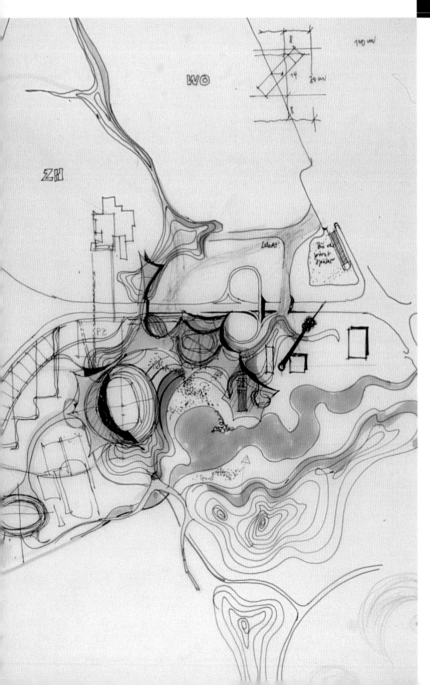

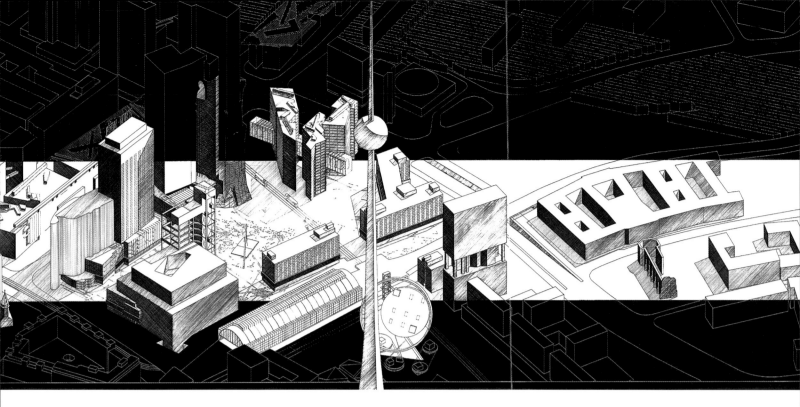

针对设计理念与想法的有效沟通以及就设计表达作出的恰当评价同项目实施过程密不可分,而这又引发了"花费多少精力参与竞赛"的问题。为了探讨这个问题,笔者采访了100位参与组织过最近10次建筑竞赛的同事,就建筑竞赛经费开销的问题向其咨询。

除了精美、逼真的透视渲染图,逻辑严密的平面图和精致的模型也是参赛者获胜的保证。他们以惊人的实力推动设计并展示成果,让业主有理由相信他们是非常值得信赖的。

在此,笔者向为本书的编写、出版作出贡献的参赛者及业主表示诚挚的谢意。尽管有些项目无法在本书中体现,但参赛者及业主对于高品质建筑环境的追求以及对设计作品的完美展示共同奠定了 [phase eins]. 工作室努力工作的坚实基础,并给予建筑师极大的驱动力。笔者同样感谢 [phase eins]. 工作室的各位同事以及众多记者。他们为"如何理解和评价方案演示成果"的调查付出了大量精力和时间,其全面、深刻的论述使本书的内容更加丰富、充实。

克里斯丁·艾希曼、本杰明·胡斯巴赫、克里斯蒂安·雷姆豪斯

竞赛随笔

建筑的瑰宝

——"映像历史":建筑竞赛及其表达

建筑竞赛如同生活本身一般丰富多彩。

1. 建筑竞赛的历史

数百年来,在建筑行业中,竞赛形式层出不穷。从文献记载的资料中很难界定建筑竞赛产生于何时、何地。建筑竞赛在建筑行业中占据重要地位已经长达500多年,如今依旧如此。尽管从16世纪到19世纪中叶建筑竞赛的评定始终缺乏具体的参照标准,但对于设计成果的不懈追求和富有争议的讨论却与今天的竞赛十分相似。然而,竞赛程序使项目委托方可以比较不同的解决方案并在决策过程中拥有更大的选择余地。通常,参赛者与业主没有固定的合作关系。建筑竞赛很可能在古代就已存在。公元前448年,雅典公民通过投票方式决定在卫城里建造一座建筑物,旨在倡导人们远离战争、珍惜和平。在文艺复兴时期,建筑竞赛迎来了它的春天。皇室和教会的贵族发起了很多次伟大的建筑竞赛;艺术家和建筑师作为参赛者,展示了诸多采用先进建筑技术的前卫设计,并提出了多项具有可操作性的实施方案。甚至有说法称,在15世纪,西班牙投资者聘请多位建筑大师为其投资项目的方案设计提出意见,然后依此进行决策。自1500年以来,意大利建筑竞赛的组织举办可追溯到文艺复兴时期对于相关问题的

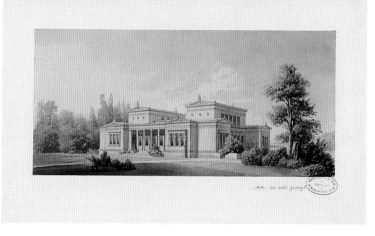

系统探讨。"以建筑竞赛促进经济发展"的想法可追溯到16世纪伊始欧洲社会受贸易影响的时期。当然，文艺复兴时期是一个集科学进步、技术创新、艺术辉煌于一体的时期，是建筑竞赛发生、发展的重要背景。据记载，1417年是建筑竞赛的萌芽之年，来自法国、德国和英格兰的建筑师齐聚佛罗伦萨来完成佛罗伦萨大教堂结构设计。实际上，15世纪的大部分建筑竞赛均在佛罗伦萨举办。大约1490年，洛伦佐二世·德·梅第奇又组织了一场佛罗伦萨大教堂立面改造竞赛。16世纪初，罗马的建筑竞赛也为世人所知，比如圣若望圣殿教堂设计竞赛。因为当时建筑竞赛的很多程序缺少成文的规定，所以结果往往不公开、不透明，甚至被出资人或赞助者干涉，导致有些方案设计的执行和最终结果饱受争议。

18世纪，相对宽松的学术氛围促进了多元化建筑竞赛流程的形成。大学、学院和其他教育机构纷纷建立并规范了建筑竞赛和评奖系统，建筑师、规划师、工程师、景观设计师甚至学生均参与其中。建筑竞赛和评奖系统产生于法国，参赛者可以参与跨国项目的建筑竞赛。1720年，皇家绘画与雕塑学会为参赛者专门设立了"罗马奖"并提供差旅费。1763年，由Czarina Katharina出资举办的圣彼得堡建筑更新设计竞赛吸引了来自不同国家、不同地域的参赛者，这是国际建筑竞赛的典型范例。此外，公众广泛参与了该竞赛的公开展览，在评委评审以及宣布获奖者名单之前人们就能看到所有提交的参赛作品，这成为当时一大亮点。

在德国，自19世纪起，建筑竞赛频繁举办。其中一项特殊的赛事是由德国建筑工程协会（Architekten und Ingenieurverein AIV）在1829年举办的"即兴竞赛"，众多参赛者在现场介绍并讨论设计作品，评审委员会由此进行评奖。1851年，为了纪念卡尔·弗雷德里克·申克尔（普鲁士建筑师、城市规划师、画家、家具及舞台设计师，德国古典主义的代表人物），"即兴竞赛"演变成一项年度主题竞赛，参赛者为德国建筑工程协会的会员；1855年，弗里德里希·威廉四世首次以国家赞助的形式资助申克尔竞赛，其由此正式晋升为一项国家级竞赛。1867年，德国建筑工程协会起草了第一部《建筑竞赛组织规范》，并获得了德国其他建筑协会的认可，经过少量的修改，最终于1868年在德国汉堡正式获批。

左页：佛罗伦萨大教堂，立面模型，1590（设计：Giovanni da Bologna）
上左图：罗马圣若望圣殿教堂，铜版画，1607（设计：Jacques Lemercier和Michelangelo）
上右图：建筑师居所，钢笔水彩画[设计：Friedrich Adler（1827—1908），申克尔竞赛1852年参赛作品]

19世纪后半叶,《建筑竞赛组织规范》开始实施,并不断地调整和修改,以适应新的要求。建筑竞赛的数量一直在稳步增长,1897年,新版的《建筑竞赛组织规范》出台。在城市化和工业化快速发展的进程中,以往从未有过的新的建筑类型应运而生,而在此之前并未建立标准模型,相应的规范在建筑竞赛进程中被制定。值得一提的是,商业领域中的银行、股票及证券交易所、百货商场和大型市场以及公共建设领域中供德国皇室和政府使用的行政大楼、宫廷建筑、法院、学校、车站、游泳馆、文化教育建筑、大型公寓住宅及城市综合体都成为建筑竞赛的内容。此外,建筑竞赛的内容还包括城市格局更新、基础设施供给(供水和排污等)、为涌入都市寻求工作机会和生存空间的大量外来人口所提供的住所,以及与建筑设计同等重要的城市公共空间设计。统计数据显示,在1868年(第一部《建筑竞赛组织规范》生效)至1889年的21年间,德国共举办了258项公平、开放的建筑竞赛,共提交了10 000余项参赛作品。

不可忽视的一点是,自19世纪起,几乎每一项有记录的建筑竞赛最终都陷入了无尽的争论,争论主要集中在"原因"和"方法"两方面。评委的构成一直饱受争议,至今依然如此。因此,也就不难解释为什么建筑竞赛的结果和趋势常常是可以预料的。然而,对于建筑竞赛最终结果的评价却很难被理解或改变,获胜者只有很小的可能性将方案设计付诸实践。众所周知,在19世纪只有半数获胜作品被真正建造出来,且经过多次调整和修改。一个著名的建筑竞赛事故即在1867年德国柏林大教堂设计竞赛中,由于遗漏了一项重要的限定条件,评审的意见出现了严重的分歧,最终评审委员会确定了新的方案设计并达成一致意见,该方案设计成为竞赛的最终结果。

1871年,在德国柏林举办的德意志国会大厦设计竞赛首次方案征集中,由于工作范围不明确,举办方在1882年决定举办一次后续竞赛。

即使到了20世纪初,仍有人抱怨参赛人数太多,尤其是并非十分重要的建筑竞赛。对参赛者进行资格审查的声音不断增多。建筑师Ludwig Hoffmann(1852—1932)在回忆录中写道,他曾担任180项建筑竞赛的评委,并在76座城市中参与评审工作。平均每项竞赛他都需要审阅100份作品,每份作品大概有10张图纸。依此合计,他作为评委一共审阅了180 000份图纸。

1903年,德国建筑师协会(Bund Deutscher Architekten BDA)成立,其致力于改进建筑竞赛组织规范。1910年,艺术家协会也被整合进来并积极参与改进工作。自此,在公共领域形成了一套完整的建筑竞赛规则体系。

一系列规模宏大且形式独特的复合性竞赛在1900年后于德国柏林组织实施。在世纪之交,德国建筑师充满了对于柏林城市发展的任意性和规模无限扩张的深切担忧。在20世纪最初的几年,德国政府牵头组织了一场以"伟大的柏林"为主题的建筑竞赛。

在建筑竞赛中，招标文件的准备成为最重要的工作。招标委员会将来自超过 176 个地区已获批建造项目的施工图加以整合，并将其纳入招标文件中。

建筑竞赛始于 1908 年 10 月，参赛者有将近一年的时间进行设计。考虑到排水系统以及为公共活动和绿化预留的建设用地，交通运输网络成为该竞赛最重要的关注点。

由于工作量巨大，最终提交的参赛作品仅有 27 份。在对匿名作品进行评价的过程中，从评委到公众，到处充斥着激烈的争论。建筑师言辞激烈地表达对政府官员和评审委员会非客观决策的不满。在 20 世纪 20 年代，该竞赛的最终结果对几个大型公共广场产生了深远的影响，如亚历山大广场、波茨坦广场、莱比锡广场。一战之后，建筑竞赛的改革重新开始，但直到 1927 年，在德国建筑师协会内部才达成共识，并制定了一系列新的规范，例如，提高协会对参赛项目的影响力、增加评委的书面说明、在已有竞赛规程中设立初审制度、协会成员必须遵守竞赛原则的承诺，以及在匿名竞赛中参赛者必须以 6 ~ 8 位数字识别号码替代当时的代码字。

一战之后，在 1921—1922 年间，一项单体建筑竞赛（尽管当时不被看好）吸引了来自世界各地的参赛者。竞赛的内容是在腓特烈大街建造一座高层建筑。该竞赛由新成立的 Turmhaus 公司主办，主题为"收购设计草图之概念竞赛"；它引发了"如何征集竞赛成果"的严肃讨论。参赛者并没有酬金，但在仅仅 6 个月的设计周期中评审委员会仍然收到 144 份参赛作品。然而，奖项已提前计划授予 Bruno Möhring、Otto Kohtz、Hans Kraffert 等几位建筑师。除了这个提前内定的结果，竞赛本身的目的和令人费解的评审结果共同引发了业内的巨大争议。被指定的评审委员会的成员有 Hermann Billing、Hermann Hahn、Ludwig Hoffmann、Heinrich Straumer 和 Paul Wittig，排在最后的评委与 Turmhaus 公司有着十分密切的关系，同时也是竞赛主办成员之一。评委的知名度直接影响参赛者对竞赛的关注。最终，一等奖授予了来自卡塞尔的 Alfons Becker、J.Brahm and R.Kasteleiner，他们在建筑历史上并不知名。密斯·凡·德罗的设计几乎没有受到评委的关注，他设计的玻璃摩天楼在世界范围内所收获的声望是通过他自己的宣传及赛后建筑评论报道获得的；二等奖授予了一个来自德国柏林的建筑团队，其成员为 Hans、Wassili Luckhardt 和 Franz Hoffmann。然而，令人遗憾的是，这些获奖方案均未实施。

左页左图：德国柏林大教堂，钢笔铅笔水彩画，1886（设计：Friedrich Adler）
左页右图：德意志国会大厦，钢笔水彩画，1882（设计：Busse 和 Schwechten）
上图："伟大的柏林"建筑竞赛，诺伊尔歌剧厅广场，铅笔水彩画，1910（设计：Eberstadt、Möhring 和 Petersen）

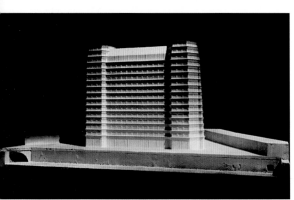

如果有人统计在 1926—1932 年间 Bauwelt 杂志刊登过的德国建筑竞赛，那么会发现有多达 643 项赛事，这个数字是惊人的。Bauwelt 杂志列出了担任评委次数最多的一些专家，其中 Ludwig Hoffman、Fritz Schumacher、Emil Fahrenkamp、Heinrich Straumer 和 Wilhelm Kreis 位列其中。

自 1933 年起，大型公开竞赛显著减少，但有大量建筑师积极参与以"概念性建设"为主题的竞赛，例如 Reichsführerschulen 和 Häuser der Arbeit。

对于大城市中异常拥挤的居住状态的反思和对于城市居住形式的深层次探索被列入建筑竞赛的主题中。一些大型公司，如埃森市的 Krupp 公司，举办了一些以"工人居住区"为主题的开放型竞赛。

在魏玛共和国时期（1919—1933），大多数德国建筑竞赛都被限定在小型廉价预制住房的发展框架下。住房协会、社会房屋企业和建设协会如雨后春笋般兴起。除了大型居住区的建造，独户住宅也被纳入建造计划。Bauwelt 杂志就是小型住宅竞赛的先驱组织者。德国柏林博览贸易交通管理局在 1932 年主办的"增量住房竞赛"是关于小型廉价预制住房的知名竞赛之一。尽管原型住宅为建筑展览而建造，但这种住房却没有普及。自 1934 年起，建筑竞赛的程序发生了相当大的改变。新程序由 Reichskunstkammer 主席 Eugen Hönig 博士颁布生效。自 1937 年 Albert Speer 被任命为民用建筑检察员起到二战结束，德国一直没有举办过公开的建筑竞赛。建筑竞赛的程序并没有被完全弃用，但所有已生效的规则经常被忽视。客观因素严重束缚着建筑师进行方案创作，完成特定的设计任务被认为是理所当然的。

2. 建筑竞赛的表达方式

在过去的 200 年中，只有为数不多的建筑竞赛任务书对图纸画法和模型的表达方式有所要求。1900 年以后，绘图技术明显改进，绘图技巧日益受到关注。一战时期，部分提交的参赛作品采用钢笔水彩的形式绘制颇具真实感的效果图和透视图。一战之后，建筑竞赛的表达方式有了明显的变化。富有表现力的炭笔图和粉笔画被用于图面表达，绘画表达因此颇具戏剧效果。

建筑竞赛表达方式的变化不仅增强了建筑的纪念效果，同时也使其更富诗意。例如，汉斯·夏隆在 20 世纪 20 年代初用水彩、木炭和粉笔将其设计融入具有灵动之光的氛围中，大大增强了建筑的表现力。其他建筑师如 Dominikus Böhm、Otto Kohtz、 Hans Poelzig、Luckhardt 兄弟以及 Alfons Anker 经常将炭笔图用于建筑竞赛中，色彩丰富的木炭画极具舞台效果，可媲美巴洛克绘画艺术。

自 20 世纪 20 年代起，一些建筑竞赛规定在图面表达中只能使用一种颜色。自此，建筑师们已经无法再用传统、古老的钢笔水彩技法展示方案设计了。

在 1921—1922 年间德国柏林腓特烈大街举办的"高层建筑竞赛"的任务书中规定，所有图面必须用线条表达。这项含糊不清的规定导致大多数参赛者提交的作品都采用炭笔制图，只有少量的人选用钢笔墨水画。举办于 1927 年的国际联盟大楼（坐落于日内瓦）建筑竞赛明确规定参赛者必须使用钢笔墨水绘图，于是绘图方式成为评奖的决定性因素。勒·柯布西耶和皮埃尔·让纳雷很喜欢该竞赛，他们提交了设计蓝图。一位保守的评委试图拒收这件作品，仅仅因为它不符合竞赛成果提交的规定；事实上，也的确如此，所以结果令人遗憾。在 1929 年举办的德国柏林交通运输公司办公楼建筑设计邀请赛中，一位评委明确表示小型竞赛中的"匿名系统"毫无意义，因为优秀的建筑师能够轻易地从图中辨认出设计者。在之前提到的 1934 年《Reichskunstkammer 章程》中，建筑竞赛的规程已包含对图纸表达所作的相关规定。参赛作品应以原作的色彩"原样呈现"并保证其质量；对于既有彩色又有黑白单色表达方式的方案设计，以单色表达的彩色插图也应完整提交。然而，"原作的色彩"究竟有何含义？这就如同无法解码的技术说明一般扑朔迷离。

左页上图：德国柏林腓特烈大街"高层建筑竞赛"之参赛提交的炭笔描图纸，1922（设计：Hugo Häring）
左页下图：德国柏林腓特烈大街"高层建筑竞赛"之赛后重新制作的模型，1922（设计：Hans、Wassili Luckhardt 和 Franz Hoffmann）
上图：参赛模型之"我们的太阳、大气和房子"，1932（设计：Hugo Häring）
下图：参赛作品之德国柏林一座邮政办公楼，彩铅水彩图，1920—1921（设计：Hans Scharoun）

在二战之后的德国，建筑竞赛再一次成为焦点。原本在1927年实行的建筑竞赛指导方针分别在1952年和1954年扩充了内容。德国的政治被动角色在建筑设计及其表达方式中均有明显的体现。战前和战时的纪念性内容在建筑设计中已经彻底消失。不仅建筑设计本身格外谦逊，就连绘图技巧也悄然改变。从20世纪五六十年代开始，大部分建筑竞赛图纸都是用细钢笔和淡彩铅笔绘制的，而这一现象在德意志联邦共和国和德意志民主共和国的建筑竞赛方案中均有体现。自1960年起，成型的设计元素，如细部绘图模板和由Letraset公司生产的干式转印刻字，丰富了建筑竞赛的绘图形式。20世纪90年代，随着计算机技术的兴起，设计技巧也逐渐全球化。通过作品的绘图方式已很难辨认出设计者，仅靠绘图技法更难以识别。近年来，已经很少有以徒手绘图方式提交的参赛作品了，只有少量建筑师将电脑渲染图用手绘方式进行部分修饰。其实，早在几十年前，匿名化的设计方式就已经不知不觉地进入国际竞赛领域，自此，个性化的图像式签名已经消失殆尽。

如今，图纸早已不是经过建筑师的双手呈送，而是通过鼠标点击电脑完成。那么值得怀疑的是，未来的建筑竞赛作品中会有多少艺术化的个人签名式建筑图纸？电脑硬盘是否会成为未来的"档案馆"？那些未被付诸实践的竞赛概念因为留存在纸上而成为历史，但正因如此，它们才具有极高的研究价值，并不应被世人所遗忘。它们反映了一个特定建筑项目的发展趋势和它所处的时代，更重要的是，只有在时间的考验和不断的对比中后人才能辨识建筑设计的品质。因此，"档案馆"是未建成方案的守护之地，更是无数未能实现的建筑思想的永恒宝库。

Eva-Maria Barkhofen 教授
德国柏林艺术学院建筑档案馆馆长

左页，从左上图到右下图：
竞赛方案之德国国会大厦扩建及共和国广场改造，炭笔描图纸，1929 [设计：Hans Poelzig (1869—1936)]
竞赛方案之德国柏林亚历山大广场，炭笔描图纸，1928（设计：Luckhardt 和 Anker）
竞赛方案之德国柏林 Rund um den 动物园，钢笔和彩铅描图纸，1948（设计：Paul Baumgarten）
竞赛方案之德国柏林 Stalinallee，炭笔描图纸，1951（设计：Hermann Henselmann）
竞赛方案之德国卡塞尔国家剧院，铅笔和彩铅描图纸，1953—1954（设计：Hans Scharoun）
竞赛方案之德国柏林东部一座新建学院，钢笔描图纸（设计：Rudolf Weißer）
上图：竞赛方案之德国埃森矿业同盟8号转换大厅，计算机打印图纸，1999（设计：Szyszkowitz+Kowalski 建筑事务所）

参赛成本

参与竞赛同样意味着经济投资。编制一套完整的竞赛展示文件无异于编写一张采购清单。这其中所付出的努力和所获得的收益只能通过中标的可能性和签订合同而获得的利润来大致估算。有的人认为，无论竞赛的奖金（或补偿金）有多少，都抵不过完成竞赛方案所付出的劳动，因此这种估算似乎是多余的。在大多数情况下，参赛者得不到一分钱，这被看作是一种经济资源的浪费，尤其是在开放型竞赛中。但尽管如此，参赛者仍然对参与竞赛满怀期待。正如同：一位医生、律师或者画家和同伴们参与一场竞赛并免费提供部分重要服务，随后却被告知主办方经过深思熟虑后签约的机会给了别人。这种看似悲惨的安排方式，是由于只有竞赛获胜者才能签约，而这是公认的标准。至于其是否公平或是否合适，则需要在竞赛体系之外进行专门探讨。

除这一基本问题外，还有很多值得思考的细节——工作并不总是公平、公正的，奖金和签约机会也是如此。竞赛主办方为举办竞赛而进行的投资也不应被忽视，并应对其严格审查。具体来说，以下四个方面应该被充分考虑：

（1）竞赛任务本身所要求的工作以及图纸绘制和模型制作；
（2）通过竞赛奖金或参赛酬金给予参赛者补偿的水平；
（3）所要求工作的内容及深度（如图纸和模型的数量等）；
（4）对于建筑服务部分的报酬和业务购置行为（非报酬性质）的区分。

1. 成本

成本是参赛者必须提前投入的部分。无论竞赛表达还是质量评价，抑或其本身具备的传达某种思想的可能性，所做的努力均不应脱离既定目标。随着模型或效果图制作服务外包，很难将图纸和模型中隐藏的技术成本与竞赛概念的生成成本分离开来，因为它们在项目进程中早已融为一体。对于大多数参赛者而言，投入多少成本只是其参与竞赛在经济负担方面的客观条件，他们不会考虑其他方面的因素，例如，提高公众形象，提升个人或团队的设计能力和专业技能。在业务层面，参赛者在参赛前必须作出决定：要么参与一个通过竞赛中标而获得经济回报的竞赛，要么提前计划并以其他经济来源平衡收支。

完成一个参赛作品所需工作的第一决定性因素是项目本身，即它的规模和复杂程度。项目是学校还是医院，是牧场小屋还是遗产地，等等，这些都会影响对项目的理解、项目概念的产生以及项目程序的运行。

为了更好地了解大家关于"竞赛方案及其表达方式所需的合理投资"的看法，笔者调查、采访了100位参与过由[phase eins].工作室组织的10项竞赛的业内同行，整理、汇总了投入竞赛的各种成本，其中也包括外包类工作。共有77位业内同行提供了相关信息。调查统计结果见第25页的图表。如该图表所示，针对不同的竞赛，参赛者的资金投入差异巨大。然而，所有参赛者投入"正常"规模的竞赛（例如，中等规模的单体竞赛）中的平均成本与竞赛总成本呈正相关关系。所有建筑师投入竞赛的总成本是竞赛总成本的3～4倍。无论这是否说明了对于这种畸形状态的高度容忍或接受，这至少体现了建筑师对于竞赛的投入程度之深。尽管这是一个小型调查，但它却明显反映了小型项目虽然投入的成本相对较少，但相比较竞赛奖金，对建筑师却更为不利，即按标准的计算方式(即通过竞赛奖金或参赛酬金给予参赛者补偿的水平）所计算出的比例太低。对于竞赛主办方而言，参赛者所提供的设计服务的总价值与设计费之间的悬殊比例说明相比较竞赛奖金的支出，主办方所获得的回报是十分巨大的。参赛者工作量的总和与竞赛总成本（竞赛方案的制作费用，付给评审委员会、组织者的报酬，等等）的比例表明，参赛者的总成本一般比举办竞赛的总成本要高。事实上也的确如此，因为即使没有竞赛，这些花销也会产生在项目进程中。

调查结果突出显示了设计团队参与双阶段竞赛和总体规划竞赛所要做的额外工作。《德国RPW竞赛策划指南2013》指出：对于同一个项目，参与双阶段竞赛的工作量是参与单一阶段竞赛的工作量的1.5～2倍。就项目特点和成果提交而言，跨学科"设计-投标-建造"类竞赛的要求更高。调查结果说明，如果提供此类服务（工程类、技术类等）应该被给予奖励，那么所有学科的费用均应考虑在内。竞赛总成本的上涨是一个客观事实，尤其是高额的建筑服务工程费用也被纳入其中，即使这些花销只有少部分用于方案创作。建筑师从更高额的竞赛费用中获益，因为竞赛主办方认同建筑师较之原来应该在竞赛进程中获得更多的收益。

下图：竞赛方案的制作费用与竞赛总成本之比较（调查对象：参与过10项竞赛的100位业内同行）

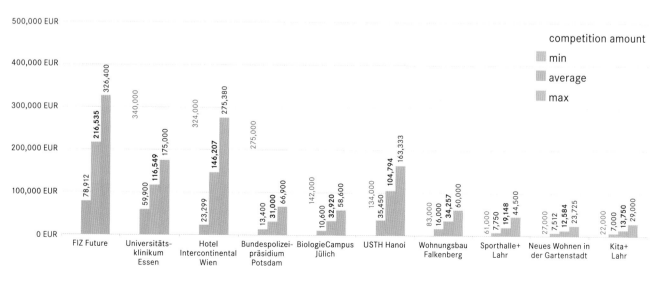

2. 竞赛奖金与成本的计算

竞赛总成本是一方面，竞赛主办方以奖金形式的经济报酬是另一方面。参赛者的投资成本及所获得的补偿程度，不同的竞赛情况迥异。

国际上，不同的竞赛体系往往决定着竞赛总成本的设定。竞赛规程的制定一般十分模糊，例如《UIA 指导文件》中的意见指出，奖金应该与项目和竞赛所需提交的内容相关联，并且所有受邀的参赛者需另外支付专业费用（参见《UIA 指导文件》中关于国际建筑与城镇设计竞赛的条例第 19 条、第 22 条，2000 年 1 月版）。英国皇家建筑师协会（RIBA）的规程也同样模糊。在这方面，德国、奥地利和瑞士的规定却相当严格。

《德国 RPW 竞赛策划指南 2013》规定，竞赛总成本至少应该与通常情况下竞赛所提供服务的费用相当。该规定表明除了设计研发的费用（占总费用的 7%），还要加上可能存在的专项规划费用。Trentino-Alto Adige/Südtirol（意大利语）自治区的《竞赛规程》也做了相似的规定。例如，《竞赛规程》第 9.1 条规定："奖金总和由以下几部分按比例分配的专项规划费用（基准收费标准）组成，其中，一等奖占 35%，二等奖占 25%，三等奖占 15%，提名奖（或荣誉奖）占 25%。"针对这一计算方法，《奥地利建筑法规》（Wettbewerbsstandard Architektur, WSA 2010）和《瑞士竞赛奖金计算标准》（Wegleitung zur Ordnung, SIA 142）分别提出了另一种解决办法。《德国 RPW 竞赛策划指南 2013》规定，竞赛总成本应由竞赛补偿费用与专项规划费用构成，二者之比为 1:1；《奥地利建筑法规》采用 4:1 的比例；《瑞士竞赛奖金计算标准》采用 2:1 的比例。奥地利的 Bundes-kammer der Architekten und Ingenieurkonsulenten 提供了一种联网式的计算工具（www.arching.at/preisgeldrechner），该工具考虑了项目的复杂程度。

值得一提的是，《瑞士竞赛奖金计算标准》更加明确地规定了如何支付与竞赛相关的附加服务费（如透视图、细部构造图、精确的预算书、专业规划整合等）及其所占竞赛所需额外支付费用的百分比。该规定确保了如果参加一项提供 100% 设计服务的竞赛，那么将获得与签订合同所得到的报酬相当的奖金。预订合同的局限性表现在与竞赛相关的附加服务费的收取上。在这方面，《德国 RPW 竞赛策划指南 2013》与之相差甚远。

《北欧国家联合规范》（北欧国家的建筑竞赛规则，§10）以工作范围与竞赛内容为基础计算竞赛总成本。然而，这种计算方式在某种程度上也是模糊的，经常要视具体情况而定，而且评委有降低总成本的自由，尤其是当竞赛成果质量无法保证或奖金不足之时。

放眼欧洲以外的地区，有一些更明确、更具体的规定。例如，印度在《竞赛规程》里对于竞赛总成本的计算方式：参考项目的规模，但并不对项目的复杂程度进行区分；对于招标项目的奖金和专项规划费用总和有着非常全面的定义。

3. 竞赛成果提交

除了上述成本组成部分，第三个因素是参赛者需要针对竞赛内容设定具体的设计细节。除了技术层面的内容（尤其是跨学科竞赛）等基本问题外，图纸和模型（包括渲染图、报表和所需的其他文件等）的数量、比例也至关重要，它们影响着参赛者的工作时间和工作内容，如模型制作、咨询服务和效果图制作等。从本质上说，方案设计阶段服务的常规标准决定了竞赛所期望的设计深度。

如果在国际范围内做个比较，便会发现，各个竞赛的规则区别巨大（见图表）。通过与竞赛主办方的讨论，评审委员会重点考虑提交方案的可行性，并把可行的方案圈画出来。相对于成果要求，竞赛费用合适吗？提交 4 张、6 张、8 张甚至 14 张 A0 图纸是否足以表达所有方案内容？绘制多大比例的总平面图、平面图以及立面图才能展现所有设计细节以及整体关系？所有楼层平面需要统一比例吗？模型应该（或能够）做得多大？一个工作模型是否足以说明问题？如果过半数参赛者提交更精细的模型，那么情况如何？一张 A3 草图是否足以表达设计内涵？是否应在竞赛第一阶段完成所有平面图或功能分区图，以便验证空间布局及建造成本是否合理？

从长远来讲，在建项目的巨额经费和无休止的延期，迫使许多业主变得焦虑并时常需要针对某个问题作出决策。因此，业主迫切需要综合性信息，公众也对越来越多的综合性图解感兴趣。于是，更多的附加分析图出现了，它们从更加易懂的层面解释方案设计。

经验证明，评审无捷径可走。评委只能凭借对提交作品的主观感觉，从方案本身出发作出决策。最基本且最重要的原则是，评委必须尽可能地了解方案的全部信息，以便通过其核心部分对方案作出综合性评价，而非单纯地依赖大量图纸。这便需要做到两个方面：第一，限制工作量，减少"无用功"，降低或抑制对方案并非至关重要的服务需求，但这经常与一些缺少安全感或需要一定保障的业主的需求背道而驰；第二，尤其针对双阶段竞赛，确保方案中比较重要的设计已在竞赛第一阶段被权衡、考虑，以便聚焦当前的工作与精力。

如果工作量需要减少，则应该在结构层面上减少提交成果的数量。例如，仅绘制单个楼层平面图，以较小的比例绘制图纸，减少印图的开销，在绘图技术方面将工作量减到最小。在计算工作量时，平面图被想当然地一直保留着，但它在评委进行评价或作出决策时难以体现出来。与此同时，与减少工作量相关的问题还包括：与概念草图配套的图纸大量减少（在双阶段竞赛中，图纸绘制完成于竞赛第一阶段）；技术顾问何时介入（例如，竞赛第二阶段）以及附加服务的内容和深入程度（效果图、模型、施工细节、预算估算和可持续性评估等）。

下图：参赛者的平均成本与竞赛程序成本及方案预算之比较（调查对象：参与过 10 项竞赛的 100 位业内同行）

Project	Number of competition phases	Number of participants (1st/2nd phase)	Average expense for participants[1] in EUR	Total expense for all participants[2] in EUR	Competition sum in EUR	Contract value[3] (approx.) in EUR	Total competition costs[4] in EUR	Participant cost/ competition sum	Total competition costs/expense for participants	Average participant cost/contract value
BMW FIZ Future	2	12/6	216,535	1,753,936	498,000	1,500,000	1,300,000	352.2%	74.1%	14.44%
Gymnasium+ Lahr	1	15	19,148	287,220	61,000	400,000	200,000	470.5%	69.6%	4.79%
Day Care+ Lahr	1	11	13,750	151,250	22,000	140,000	145,000	687.5%	95.9%	9.82%
BiologieCampus Jülich	1	13	32,920	427,958	142,000	800,000	380,000	301.4%	88.8%	4.11%
USTH Hanoi[5]	1	6	104,794	628,767	134,000	5,960,000	450,000	469.2%	71.6%	1.76%
Federal Police Headquarters[5]	1	23	31,000	620,004	275,000	1,690,000	450,000	225.5%	72.6%	1.83%
University Hospital Essen[6]	2	16/6	116,549	1,107,211	340,000	9,450,000	640,000	325.7%	57.8%	1.23%
New Housing Falkenberg[5]	1	8	34,257	274,057	83,000	370,000	220,000	330.2%	80.3%	9.26%
New Living in the Gartenstadt	1	7	12,584	88,087	27,000	140,000	100,000	326.2%	113.5%	8.99%
Hotel InterContinental Vienna[7]	2	24/6	146,207	1,798,348	324,000	4,400,000	620,000	555.0%	34.5%	3.32%
Average								404.4%	75.9%	6.0%

(1) Participant data varies taking into consideration overhead costs and hourly rates of owner/partners.
(2) The average value was taken for participants who did not submit data to the survey. In two-phase competitions the assumed cost for participants who only submitted to the 1st phase equals 35 % of the average cost incurred for participating in both phases.
(3) Approximated value is based on the contract volume to be awarded through the competition (i.e., not the entire contract, but assuming: fee zone III, average rate following HOAI or similar, without inclusion of ancillary costs). In cases of chargeable costs greater than those listed in the HOAI tables, extrapolation based on the Richtlinien für die Beteiligung freiberuflich Tätiger (RifT) is assumed. For the BMW FIZ Future competition the contract value includes only production of the masterplan, no building planning.
(4) Sum of prize money, fees, jury fees, procedural costs for model building, printing etc., travel expenses and competition management
(5) Including landscape planning
(6) General planner contract
(7) Approximated contract volume for schematic design, preliminary planning and building application, as well as specification details and artistic direction

Lieber Benno,

hier der ein A4 Vorschlag zur ersten Phase im Wettbewerb für die Moschee in Köln. Verfasser unbekannt. Flog leider in der 1. Phase raus. Die Arbeit war ohne Zweifel lesenswert und gerade was das Städtebauliche Konzept und die gestalterische Absicht anbetraf völlig ausreichend.

Mit Overhead + Originalbetrachtung auf dem Tisch völlig ausreichend, (allerdings nicht mehr bei 20-30 Personen).

Unsere Verfahren stimmen hinten + vorne nicht mehr, weil zu viel widersprüchliche Interessen zusammengebacken wurden.

Gruß Max

上图：Max Bächer [德国科隆 Diyanet Isleri Türk Islam Birligi (DITIB) Central Mosque 建筑竞赛（举办于 2005 年 11 月）评审委员会主席] 写给本杰明·胡斯巴赫（ [phase eins]. 工作室成员之一）的一封信

亲爱的本杰明·胡斯巴赫：

这是即将举行的德国科隆 Diyanet Isleri Türk Islam Birligi (DITIB) Central Mosque 建筑竞赛第一阶段的大致方案，请您过目。

竞赛内容比较广泛，涉及建筑设计以及城市概念规划等相关领域。经过充分的研究，竞赛主办方为各参赛者及其团队准备了一笔供 20～30 名成员使用的经费，作为参与竞赛坚实的物质保障。

竞赛的程序设定比较复杂，需要权衡各方利益，但我们会竭尽全力做好准备与服务工作。其间若有不周之处，还请您批评指正。

非常感谢。

Max Bächer

右页：德国科隆 Diyanet Isleri Türk Islam Birligi (DITIB) Central Mosque 建筑竞赛（举办于 2005 年 11 月）第一阶段方案图纸（设计：Kister Scheithauer Gross）

几乎在所有竞赛中，模型都是设计工作不可或缺的部分；它不仅推动设计顺利进行，还是评委讨论并评选最佳方案的依据和载体。作为减少参赛者工作量的一种方式，在某些特殊情况（尤其是当竞赛经费不足之时）下也可考虑"省略"模型。模型的比例和精致程度对模型制作及运输费用具有很大影响，而这些是可以节省的潜在费用。模型的比例和整体尺寸应该恰到好处，没有一丝一毫的"浪费"，衡量标准即模型能够体现设计之美和表达空间效果，这也是模型相较二维图纸的价值所在。对于建筑竞赛而言，1:500 的比例一般是可以接受的；对于城市规划竞赛而言，1:1000 的比例比较合适。如果需要更加精细的成果，则应适当地增加专业费用。

关于三维表达方式，其效果和实用性的合理关系众说纷纭，这不仅仅因为其花销。制作一张渲染图是一笔不菲的花销。通常，制作一张渲染图要花费 1000～2500 欧元，具体取决于图纸的大小和数量。因此，每项竞赛都应充分考虑渲染图的制作费用，并提高标底费。不容忽视的是，制作渲染图需要建筑师额外的准备工作，整体工作量甚至更多。然而，如果"制作渲染图"被省略，那么节省的花销会很多。抛开这些因素，一些参赛者希望以最佳方式展示成果、交流想法，很多业主也希望通过渲染图更深入地理解方案，并在竞赛之后与公众交流。渲染图则在这方面起到了决定性的作用。一些反对的意见要求省略渲染图或至少减少渲染图的数量并降低其重要性，其理由是渲染图虽然可以烘托氛围，但缺乏真实感并对决策过程的客观性造成一定影响。

也许这看起来有些过时，但模型仍然是非常重要的设计工具。好的模型可以展示设计推进过程、城市关系和空间序列。它们应当被给予更多的关注并以可能的最大比例加以展示。建筑师通过模型进行设计，为何不把它强调为一种基本工具呢？

Frank Barkow 和 Regine Leibinger

总的来说，在这个问题上同样需要找到一个"项目与状态相关"的解决办法。

限制对于质量的期望值，例如"简单而必需的体量模型"或"一个简单、图形化、抽象、线稿绘制的三维体量图"，等等，这些只有在描述清晰的前提下才是可靠的，这一般会发生在学术讨论会上关于一些如"评委如何看待那些超出常规工作的提交作品"的必要的讨论中。在与三维制作的精细模型比较时，评委如何评价"简单的体量模型"，如何对待带有额外三维表现的提交作品？面对设计和绘图技巧的快速发展，原先的竞赛规则已很难继续保持下去。

当需要提交比常规要求更多的内容时，德国、瑞士和奥地利的竞赛规则中有不同程度的"关于哪些竞赛成果需要增加"的建议。

《德国 RPW 竞赛策划指南 2013》规定，合理的竞赛成果的增加，应该在所需工作量超过该规定中附 II 条的相关规定时才可实施。然而，该规定中没有关于增加的百分比或类似说明。《奥地利建筑法规》和《瑞士竞赛奖金计算标准》则对相关要求做了更明确的规定。

在一定程度上，这些说明和规定是为了防止提交作品具有太过宽泛的主题。然而，在不同的尺度下做符合实际的工作，或更确切地说，所需的工作只通过一系列普遍、有效的规则来定义，因此，这需要每个人为自己的行为负责。通常，对于高估渲染图的表现效果，而忽视一些更加重要且更需注意的工作，例如预算估算和可持续性评估等。当需要扩充提交成果的内容时，所有参赛者均应得到报酬，以保证这些额外的款项公平地被每位参赛者分享，而不仅仅是竞赛获奖者。

尽管如此，还是有很多竞赛主办方希望在竞赛阶段就得到全面的方案细节信息，尤其是在具体预算方面，然而这往往很难实现。无论竞赛方案的制作费用预估设在 DIN276 的第一层级是否有意义，增加的付费服务仍有待观察。如果与建筑师的合同是基于竞赛结果协商签订的，并且建造的最高造价在合同中也是一项内容（在保证质量和原定计划的情况下进行付费），那么在竞赛进程和协商过程中就需要对预算进行全面的考虑。另一方面，如果提交的表现图在赛后马上用于公关，那么在这一阶段也需要有合适的表现图。附加费用一般为 5%～10%，付费服务应该根据个案的具体情况而定。因此，对附加费用的担忧是可以理解的。

4. 以竞赛促进经济收益的成本 - 效益分析

合理地区分建筑设计付费服务和非付费服务在现实中比想象中要困难得多，因为提供或接受一项服务作为一项具体行为，有可能导致一份合同的最终签订。从综合的法理学角度考虑，建筑师的大量活动可专注于以合同为起始的目标，而不必在规定的职业费用构架中纠结于最低费率。因此，在建筑服务类竞赛中没有严格地遵守薪酬竞争规则是比较常见的现象，只要获胜者能够真正获得建筑服务项目合同。然而，对规划服务的评价却更加困难，因为其不包含在评奖流程框架中（如《德国 RPW 竞赛策划指南 2013》的相关规定）。在这种情况下，规划服务被假定属于竞赛成果提交的范围。基于此，从司法角度考虑，以专项规划框架中的服务费计算完整的酬金是不恰当的，而应该将其仅仅作为工作量的一部分。那么下一个问题是，多少费用才算合理？

右页：Wolf D.Prix 在竞赛过程中正在做方案陈述

在计算基于工作量的补偿金之前，应重点讨论这些补偿金是否真正补偿了所产生的工作量，或是否需要预先支付少量补偿金。因为，通常情况下，一个建筑工作室，与其他公司一样，每年要有固定的经济收入才能维持运转。奖金和补偿金（对于生产服务的补偿金）往往无法冲抵全部花销，而仅仅相当于其中一部分，所以，实际的生产成本经常超出预算。

那么，预算到底有多少？想要得到确切的答案，必须仔细研究每个建筑工作室的数据统计，然而，研究的结果常常仅能得到一个笼统的数字。比得到一个具体的数字或百分比更重要的是，这个笼统的数字说明，每个建筑工作室的预算计划是非常重要的，因为方案运作全过程所产生的费用以及竞赛奖金均被纳入其中。只有基于这一视角，才能厘清成本和利润之间的关系，进而形成关键性的数据并从中找到制订预算计划的参考标准。

笔者在对相关的预算资料进行调查、整理并采访了一些业内同行后，仍然无法得到确切的答案。在这方面情况各异：某些建筑工作室，缺少成本开销的日常记录，未形成系统化的数据统计，没有预算计划；多数建筑工作室将上年度收入的 5% ~ 10% 作为生产预算。对于大型建筑工作室而言，这部分比例会降低；对于小型建筑工作室而言，这部分比例会升高。一方面，建筑师应该清楚地计算其个人及其所在工作室参与竞赛所产生的经济成本、投入一系列工作所用的时间以及参赛作品所应包含的内容，等等，以便让局外人更加全面地了解项目方案。另一方面，建筑师应该尝试尽量全面、客观地评价竞赛，并尽最大努力获得签约的机会，例如，竞赛是开放的、受限的还是邀请性的？评审委员会由哪些专家组成？参赛者的个人能力与竞赛任务及目标之间有何关系？项目方案有多大规模？等等。

建筑师与竞赛

如果现在有人认为赢得竞赛的机会是均等的并且只有最佳方案才能取胜，那他不如去买一张 6 欧元的彩票，那样获胜的概率会更高（或更公平）。

或者，你能想象 120 位、800 位甚至 3000 位外科医生在手术台前参与一项竞赛，比试谁能做"最好"的外科手术，并且他们自己还要掏钱来参与竞赛？从来没有！

Wolf D.Prix/COOP HIMMELB(L)AU

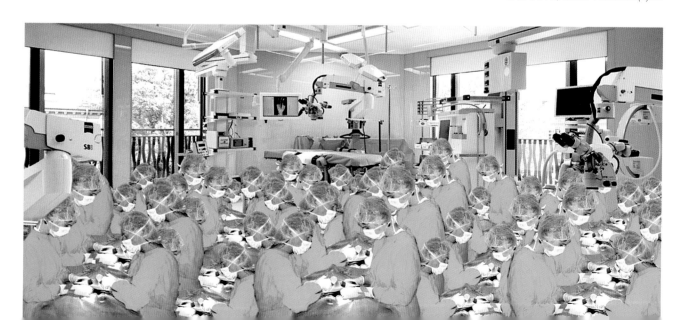

小结

考虑到各类复杂的因素和不同的利益，很难作出一个全面、有效的总结，不过在某些方面可以形成结论。

（1）减少工作量。
参赛者可以通过两项措施减少工作量：在进行基本工作之前提交概念表达的相关内容，或放弃提交超出标准工作范围的成果。例如，前者可广泛地推测和评估哪个方案在这一设计阶段价值有限，但前提是，推测和评估不能改变方案设计原有的表达深度，并且不能增加评委评审的难度。公共关系方面的工作也应以竞赛为根据。一般来说，可以在双阶段竞赛第一阶段减少概念表达的提交内容，这将大大改变竞赛进程。省略平面图、立面图或剖面图以及缩小比例对减少工作量的意义并不大。

（2）高额补偿金。
在相关的竞赛规程中编入条款来计算竞赛总成本，如奥地利和瑞士的相关规定以简单、有效的方式合理地设置参赛者参与竞赛所需承担的工作量与竞赛总成本之间的关系。与《德国 RPW 竞赛策划指南 2013》相比，奥地利和瑞士的竞赛补偿金显然更高。针对小型项目需要做一些必要的调整。在实践中，计算竞赛总成本时应合理考虑附加服务的费用以及以何种方式公平、公正地将其分摊给参赛者。

毋庸置疑，竞赛推动创新。这一点已经达成共识。工作室独立自主地完成一个项目，不受任何外界的干扰，这无疑是一种特权。如果还能有机会反映现实、解决疑问、探索未知的领域、推动学科超越既有框架，则更加幸运。由此看来，竞赛是一种技术研究的方式，把工作室变为实验室或工坊。建筑师克劳德佩罗为卢浮宫东立面设计的柱廊，约翰·伍重为悉尼歌剧院设计的贝壳屋顶，等等，这些都是在建筑竞赛中形成的不朽之作。与之类似，20世纪20年代中叶日内瓦国联大厦竞赛和80年代初巴黎拉维莱特公园竞赛都证实，竞赛作为一个知识交流和技术展示的平台，对文化和价值观的形成与发展具有重要作用。建筑竞赛根植于建筑本身，并且极大地推动建筑技术乃至建筑生产方式的进步与创新。然而，凡事有利也有弊。上述这一切都要有所代价。撇开理想主义的原则，竞赛如同阿喀琉斯之踵一般成为追求职业梦想的建筑师的致命弱点，甚至可能走向衰落。这种状况已经持续相当长的时间，目前依然如此。如今，竞赛大多作为一项产业，承载着建筑师实实在在的渴望。为了一些不值得的竞赛，建筑师簇拥而上，被引诱到这个游戏中，等待他们的只是镜花水月般不确定的奖项和渺茫的获奖机会。没有任何职业会深陷至此，劳碌却最终一无所获——"彻底耗尽智慧"却难以为继。不可否认，这的确是败坏世风的。问题是如何摆脱这一困境。或许，可以改变"参与竞赛"的习惯或从既有竞赛系统中寻求新的出路。前者等同于呼吁所有人拒绝参与，而后者则需要做一些努力；最重要的是所有参赛者、主办方及其他参与者都愿意支持最为公平、公正的标准，避免任何形式的私利与剥削，并且尊重那些无偿投入大量心血却不知最终结果如何的建筑师。无论最终选择哪一条路，当前建筑竞赛的形势已危如累卵。

Marc Angelil

竞赛绘图表达

在竞赛中，参赛者和主办方的沟通一般局限于视觉表达，如平面图、文本、模型。这种现象是割裂且不正常的，但平面图、文本、模型也不失为简明扼要地提出设计想法的好工具。好的设计能够自我诠释，通过几张图片就可以表达清楚。然而，大多数情况下，每个项目背后都有一个潜在的合作伙伴，拥有强大执行力的业主和设计团队是项目获得成功的保证。为了保持良好的合作关系，相关人士应该事先见面沟通。

模型制作方面的话题在此不再赘述，因为几乎所有问题都关乎规模、材料、工具、完成度、模型技巧等，这会开启一个全新的话题。其他可能存在的媒介如电影和动画也不在讨论范围之内，尽管它们在日益现代化的竞赛中起到愈发不可替代的作用；其潜在动力与发展前景将在其他方面深入探讨。人们不必惊讶未来建筑师的日常工作中可能出现一些新形式，如三维空间虚拟演示、基地现场独立演示或其他先进的技术形式，这些都将成为竞赛成果的一部分。

今天，利用电脑技术进行图像表达只是竞赛表达发展过程中的一个阶段。然而，它们代表了竞赛背景下典型的图像表达技术，这就是本书讨论的主题。

竞赛表达的内容

参赛者所提交的作品是对竞赛任务的最佳回应。竞赛任务的构想体现了建筑设计、城市规划、风景园林及相关学科之间的一种多元化构成，竞赛表达也需要包括不同的内容。在融合相关内容的基础上，必要信息的表达需要合乎逻辑并且易于理解，其往往综合了竞赛任务、设计意图、功能需求和技术要求以及对于上位规划策略的解读。所选择的演示文稿形式必须明确竞赛任务和方案设计的特点。

• 为了充分展示方案设计的技术标准和功能布局，细节概念或多或少地需要以图示呈现，如外形、尺寸、平面组织、结构功能、与既有建筑的关系等。总平面图、平面图、剖面图、立面图、表格、平面图解、功能图解等通常最先通过二维或三维绘图来完成，并与技术图纸密切相关。

• 为了表达设计想法及其衍生概念，设计的吸引力和氛围往往借助于与可实现的视觉艺术紧密相关的图示表达，如草图、拼贴画、透视效果图、彩色原料等，同时也包括兼具描述性和解释性的文字说明。

下文中记录了不同的案例及相应内容。将每部分内容加以整合并形成完整的表达,可以作为方案设计的一项附加内容。其他成功案例见"其他竞赛项目"。

图示表达的质量

在对这些案例的解读中,笔者不希望给这些绘图贴上"好"或"坏"的标签,评价什么才是画图的正确方式,也不愿评价什么是合适的风格。相反,笔者希望这些绘图可以给建筑师带来些许启发,通过全面的理解,强化对于单项内容以及整体内容的认识。

如果想要知道什么是更好或更坏的表达(基于"适宜性"的考虑),那么需要评价整体表达是否保持平衡。好的设计是在不同的要求之间创建平衡,表达也应该在清晰和聚焦的基础上寻求平衡。完整地表达设计理念是方案设计所面临的一项挑战。显然,这需要避免不易阅读或自相矛盾的表达,尽管这在日常生活中有时难以奏效。

如同任何得以传播的概念和计划,目标受众是竞赛表达的核心。通常情况下,目标受众是评委,他们组成人数众多且学科背景各异的专家委员会,评委代表了竞赛主办方和未来的业主。因此,参赛者有必要进行换位思考,进而判断评委的喜好及其相互关系。作为陪审员、技术顾问或是预审员参与项目的评审可以收获非常宝贵的经验。

项目执行的决定很少是在评审期间完成的。方案设计最先征求大众观点或其他委员会的审查,这使公众成为愈发重要的一部分目标受众。这种需求在很多以服务和图解的方式与公众沟通的竞赛中尤为重要。《竞赛参赛者的开销》一文中称,如需额外提交成果,则应给予参赛者相应的补偿金。

基本的演示架构可以在评估目标受众的期望后再进行构建,同样在此之前应该合理估计参赛者愿意承担的竞赛投资额和希望获得的预期利润。基本的演示架构包括图纸表达的范围和细节深度(假定没有预设)。目标受众的关注度及其职业与文化背景(尤其是国际项目)都影响着参赛者对于竞赛表达总工作量的判断,包括细节深度、标注以及平面色彩等。

上图:德国柏林 Baum-schulenweg Crematorium 建筑竞赛,建筑师 Axel Schultes 和 Charlotte Frank 的方案设计,1992

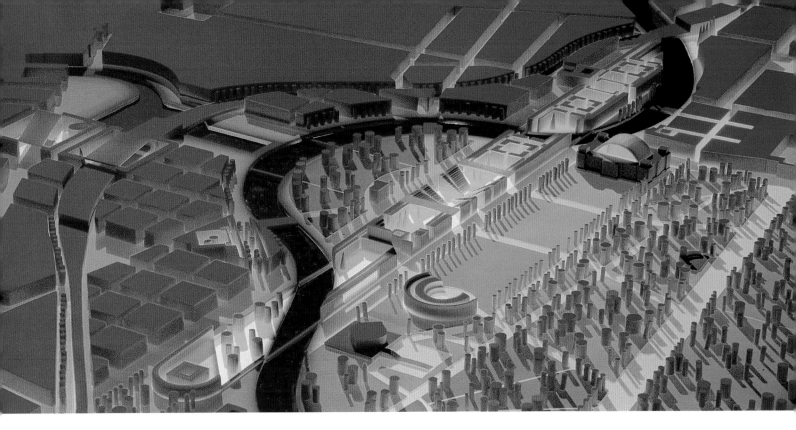

对于概念表达和设计生成，如果建筑师委曲求全地放弃基本原则，完全无条件地迎合业主不切实际的要求，那么最终结果必定差强人意。实现形式化和程序化的要求涉及另一层面的内容。为竞赛提供赞助的业主通常希望方案设计在契合设计意图及诉求的基础上，拥有创新、自主的表达方式。

总平面图和模型是方案设计重要的表达形式。鉴于当前建筑话语的保守趋势，立面绘图比图形符号更加重要。一个优秀的方案设计，其建筑的选址有特定的表达方式，二者一一对应且具有紧密的内在联系。

Markus Allmann

上图：德国柏林 Spreebogen 建筑竞赛，建筑师 Axel Schultes 和 Charlotte Frank 的方案设计，1992

通常情况下，参赛者可使用渲染图向业主和房产代理展示作品所传达的场所精神，场所精神对于业主和房产代理来说意义非凡。然而，渲染图并非总能起到决定作用的真实描绘，而是对于既有建筑关系的表达。

参赛作品需要对建筑比例、结构和体量等进行高水平且独特的解读，因为很多建筑都坐落于人口密集地区且毗邻世界著名历史文化遗址。极具表现力的平面、立面及空间印象在评审体系中至关重要。

参赛者是否理解建筑组织内涵，只能在方案提交后仔细审查。平面图的合理性、城市规划问题的解决策略和特殊功能区的重点设计等都无法通过浏览的方式发现，在竞赛中有必要对其进行评审。多次参赛的建筑师在这方面经验丰富。在竞赛过程中基于上述标准所确定的方案设计尤其会让业主甚至参赛者自己完全信服并信心满满，进而在竞赛中取得胜利。

Thorsten Schmitt, Chairman of the Berliner Bau und
Wohnungsgenossenschaft von 1892 e.G.

表达与策略

对于一个方案设计，什么才是业主或评委关注的内容？首先是清晰的表达。方案表达不应过于杂乱。此外，业主比较关注方案设计如何运作（这关乎整个场地和建筑自身的规划与设计，例如，如何规划空间交通动线，如何处理其所产生的一系列问题）。方案设计不应止步于场地或建筑本身，高质量的方案设计必须倾其全力地表达场地及建筑从内到外的衔接过渡关系。

除了清晰的表达，方案设计的哪些部分能够引起业主和评委的关注呢？极具震撼力的渲染图或极具艺术美感的标志通常会引起业主的注意，也会吸引评委花更多的时间仔细研读方案设计。亚历山大图书馆的 Snøhetta 入口和奥斯陆歌剧厅有一个共同之处，它们分别在图版上插入了具有标志性的日本国旗和捷克国旗。

Charles Moore 是一位美国建筑师兼教师，他经常作为初审员或评委参与美国加利福尼亚 Sea Ranch 建筑竞赛和德国柏林 Tegel Housing 建筑竞赛的评审工作。他曾经说过，在无数的图版中，哪怕只是滴上墨水，有时都会引起评委的注意。然而，针对方案中那些棘手的设计问题，清晰、明确且合乎逻辑的解决办法依然无可替代。

风险因素

那么问题来了，参赛者是否应该增加最初任务书中并未要求的有关设计方法或技术的内容？在这种情况下，常常会有一些"风险因素"。评委一般会更倾心于此，前提是他们认为其不会超出预算范围，这在设计进程中是可以讨论的。然而，如果方案所采用的设计方法或技术是一个前所未有的"新事物"，那么参赛者就需要非常乐于接受"新事物"的评委来接受它。

最近一个典型案例是美国匹兹堡卡耐基梅隆大学工程技术大楼竞赛。四位参赛者分别是 Zimmer Gunsul Frasca (ZGF)、Bohlin Cywinski Jackson (BCJ)、Wilson Architects 和 OFFICE 52。Zimmer Gunsul Frasca (ZGF) 和 Bohlin Cywinski Jackson (BCJ) 在至少 5 个城市里拥有工作室；位于波士顿的 Wilson Architects 已经完成了许多大型项目；位于波特兰且只有 10 人的 OFFICE 52 被认为是这些"设计大家"中持"外卡"的参赛团队，其团队中的建筑师 Isaac Campbell 认为若想赢得竞赛则方案设计必须与众不同。在竞赛中，一座七层高的建筑被建造于一片校园场地中。OFFICE 52 决定将部分功能区安排在相邻空间中，把建筑主体设为四层，以使其与基地更协调。其他三个参赛团队都遵循了竞赛任务所设定的原始要求，设计了七层高的建筑。公共基础设施设置于基地地面下，包括为整个校园供电的主要供电线路。如此一来，这座 7 层建筑就会出现一个问题，当地施工单位表示因公共基础工程建设，校园不得不关闭至少一天。通过建造四层高的建筑，OFFICE 52 成功地规避了这个问题，而这无疑对其获得项目委托起到了重要作用。四层高的建筑规避了潜在的建设问题，从美学角度为设计增色不少，也没有增加预算。这个案例充分说明，如果参赛者想要脱离竞赛任务所设定的原始要求并在众多参赛者中脱颖而出，就必须在保证方案设计高质量、与众不同的同时不会过多地超出预算范围。

Stanley Collyer, Editor-in-Chief, Competitions

我是否会被方案设计的表达形式所影响？

这是一个每个人都会否认的问题；相反，每个人会宣称"只有内容才是最重要的"。然而，通过对竞赛资料的整理，便会发现，事实并非如此。方案设计的表达方式具有至关重要甚至决定性的作用；它能够捕获观者的注意，吸引其更进一步地研读作品。每个方案设计需要一个、两个或三个夺人眼目的地方，它们能够清楚地表达方案设计的重点并激起观者的兴趣，否则极有可能陷入"被忽略"的不利境地。一旦掌握这其中的规律，观者甚至整个竞赛都会被其影响。

Thomas Hoffmann-Kuhnt, Publisher, Wettbewerbe Aktuell

开放型竞赛（100名以上参赛者，开放申请程序）与邀标型竞赛有什么区别？

这两者区别很大。从各种角度来看，我对开放型竞赛持有批评甚至反对的态度。一方面，开放型竞赛导致了时间、精力和费用的浪费。另一方面，它绕过了一些我认为很重要的事情，如与未来的业主交谈，等等；然而，了解业主的意图与诉求并与其保持和谐的信任关系恰恰是项目成功所必需的。在邀标型竞赛中，"邀标程序会减少青年建筑师的参赛机会"这个说法并不成立。例如，在法国，很多青年建筑师、刚成立的建筑工作室甚至境外建筑事务所被邀请参赛。在德国也大抵如此。

Werner Sobek

对于竞赛进程的质量而言，参赛者的数量并不重要，重要的是评委的遴选和投标文件的编制。

个人客户与公共客户只有微弱的区别。如今，公共客户或代表公众利益的审计公司越来越多地采纳个人客户的建议与做法。

竞赛的初审报告往往如同百科全书一般"兼收并蓄"，竞赛涉及的所有内容均被纳入其中，而其后果则是，重要的内容有可能被忽视，细枝末节的内容有可能被无限夸大。评委和技术顾问害怕遗漏任何事情的恐惧已超越发现新事物的渴望，各学科的艺术文化维度也不断消失。这种"喧宾夺主"情况亟待改变。

Markus Allmann

青年建筑师和刚成立的建筑工作室可以通过参与开放型竞赛来提升业务水平、锻炼演示能力，这是一件幸事。然而，有趣的是，90%的竞赛获奖者都是所谓"常客"，只有一两个新工作室能够挤进获奖者队伍中。这是因为评委太过保守还是赢得竞赛确实需要有长期的实践经验积累？

Louis Becker

在开放型竞赛中，评委不可能给予提交作品太多的关注，因为他们根本没有足够的时间深入了解方案设计。

不包含现场讲演或建筑师采访等环节的竞赛同样缺乏价值，业主和建筑师没有任何沟通，而沟通恰恰是项目成功的关键。

Ben van Berkel

在开放型竞赛中，颇具表现力的渲染图可以让一个外行对方案设计作出直观的评价。此外，开放型竞赛也是针对某些重大社会问题（如有关城市空间建设进程的民主讨论）的解决方案初步形成乃至最终落实不可或缺的推动力。

参赛者众多的开放型竞赛对于评委来说是一个挑战。为众多参赛作品评图的工作单调、乏味。只有少部分不凡之作出自青年建筑师或刚成立的建筑事务所之手；在竞赛过程中，他们经常展开有关"废旧立新"的激烈讨论。

有参赛资格限制的竞赛，往往导致专业化和单一化，不给年轻人机会，甚至对成熟的建筑师与建筑事务所也"大门紧闭"，这与开放、自由、创新的竞赛理念背道而驰。

Thomas Willemeit

需求

以业主的角度，对每个参赛作品（包括渲染图）的表达都有相同的期待是可以理解的，这是制定游戏规则的一种方法。可比性经常来自建造成本和理念表达等方面。我们都喜欢建筑师自由地表达设计理念，正所谓"你展示的就是你想要的"。

Frank Barkow 和 Regine Leibinger

认知与表达

通过总结之前多次参赛经验,我发现,准备方案演示文稿本身就是方案设计的最后步骤,包括规划结构、图解设计和插图、图片图形及文字说明、楼层平面,等等。参赛者通过一场成功的方案演示向评委展示其所认为的高质量的方案设计。就如同写一封信,字体、书写形式、排版和内容是紧密联系的,它们共同表达了信件的"精神"或"主题思想"。虽然有些特别的图解(比如,建设服务概念总是很相似)经常扮演从属的角色,但总体上,方案演示的所有内容是一体的。

Kaspar Kraemer

参赛者如何更好地表达设计想法?——通过图片、文字、图解、视频?

如今,有很多设计展示的方法。将图片、文字、图解、视频四者相结合,不失为一个好办法。除了建筑质量和功能结构的考量,每个方案设计需要建立一个平台,它就如同一个物品的使用说明。除了平面图、立面图和剖面图,这个平台可以较多地尝试运用多媒体技术。

Louis Becker

方案设计的最佳演示方法是向读者提供一个交互性工具,吸引其走进建筑师的思想世界。演示内容可以包罗万象,但必须凝练、概括地组织成一套完整的叙述,每个图像、图画或图解都在讲述一些独特的内容,并串联成一个更好的故事。

Daniel Sundlin, Partner at BIG NYC

一些获奖作品的表达方式极其简单,我们甚至无法想象这是如何实现的,例如,悉尼歌剧院竞赛、蓬皮杜中心竞赛或柏林音乐厅竞赛。约翰·伍重、理查德·罗杰斯和汉斯·夏隆在竞赛中运用图解清晰地表达设计策略。如今,大家痴迷于展示方案细节,甚至还没到概念设计阶段,逼真的表现图就已出现,其结果是,最终的质量只能大打折扣。完美的外表已经耗尽了原本应该用于探索设计理念的时间和精力。事实上,方案设计必须朝着一个具体的目标一步一步走向成熟。为了在竞赛阶段把所有内容组合成一份令人赏心悦目的演示文稿而将准备时间压缩至只有几个星期,因此,设计理念与重要的细节处理并非深思熟虑后形成的,最糟糕的是简单重复。

对于渲染图的依赖导致参赛作品的表达方式极端同质化。初出象牙塔的学生与普利茨克奖的获奖者当然是有区别的,大多数建筑工作室拥有并熟练运用以渲染图为基础且达到市场认可标准的专业三维绘图工具。因此,作为评委,应该尽量忽略渲染图,并从平面图、立面图、剖面图、图解和文字说明中寻找清晰、独创的设计理念与细节。

Matthias Sauerbruch

上图:越南河内科技大学校园竞赛,评审委员会的成员正在讨论参赛作品

图版布局

为了保证竞赛环境的公平性，并使提交作品具有一定程度的可比性，竞赛任务书中通常包括提交成果的清单和对相关表达形式的明确要求。通常情况下，参赛者必须遵守图版布局的规则，并在设计过程的前期就着手考虑图版布局的相关问题。

- 设计的重点是什么？最适宜的表达方式是什么？
- 需要花费多少精力？
- 外部设计是否有助于设计理念的表达与完善？
- 是否完全遵守图版布局的规则？

基于上述问题，图版布局不仅关乎每张图片及其排列顺序，同时也决定了图纸表达的总体构架。在某种程度上，图版布局以审美为导向，例如，大量使用全尺寸彩色图片。图版布局应注意以下几点：

- 基于重要性的不同，表达元素需要有层级划分；
- 符合竞赛任务所设定的基本要求（方位、比例尺等）；
- 距离评委 3 ~ 15 米时是否具有可读性；
- 图版编号或图标有助于指导挂图方式。

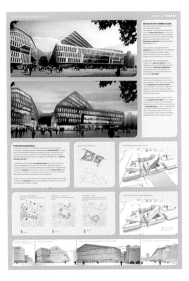
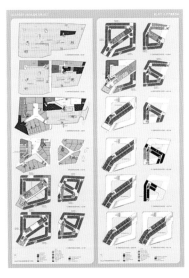
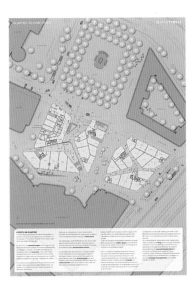
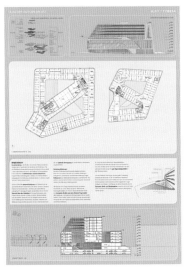
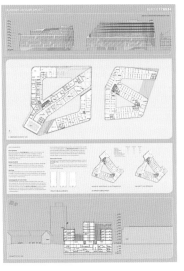
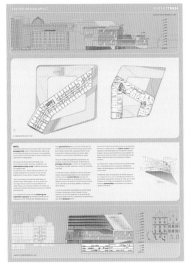
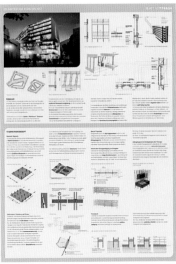

上图:德国慕尼黑 BMW FIZ Future 建筑竞赛第一阶段,评委正在对最终入围的 6 个参赛作品进行比较
左下图:德国斯图加特卡尔广场邻近地区竞赛,UNStudio 建筑事务所的方案设计
右下图:德国柏林花园城区法尔肯堡居住区竞赛,ROBERTNEUN™建筑事务所的方案设计

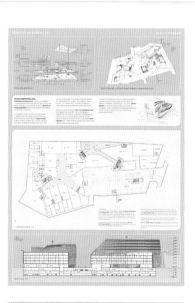

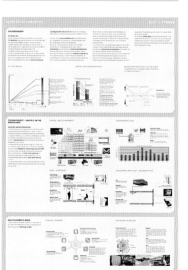

三维表达

在竞赛中，一方面，关于三维绘图表达方式的目的和优势存在一些争议，这些在很多建筑师的陈述中有所体现；另一方面，三维草图总是很迷人，大概是因为其最接近方案最初的创意。

透视图、轴测图、蒙太奇照片、拼贴画或逼真的三维渲染图以或多或少模拟现实的方式使人身临其境，直观地理解和评价空间实体。正因如此，它们是非真实的，因此备受争议。

以上这些意味着参赛者需要额外地增加工作量。除模型外，一张令人信服并迅速理解的表现图能够极大地增加其对评委的吸引力。不同的表达要求需要结合不同方面的内容：

- 视点（室外／室内，鸟瞰／透视）；
- 画法（水彩、线图、CAD、拼贴画等）；
- 色彩与材质；
- 内容（景观元素、配景人）；
- 气氛（日景、夜景、天气、光线等）。

这些表现图在几十年前很少出现，而在如今的竞赛中却不可或缺。

当前，参赛者面临的挑战是如何在全球化影响下展示个人风格。如今，图像和效果图绘制呈现彼此趋同的潮流，最近已发展成一种可与建筑模型制作相媲美的产业。熟练掌握并灵活运用图像和效果图的绘制技术意味着参赛者能够在电脑上创造图形杰作。然而，另一方面，渲染图也因其潜在的欺骗性而一直备受争议。

在某种意义上，建筑是情感的集合，此即渲染图存在的理由。目前，渲染图仅仅停留在"艺术的状态"，这是亟待改善的。
Jakob Dunkl

渲染图十分重要，尤其对于一些外行评委而言。然而遗憾的是，大多数提交作品的渲染图的艺术水准相对较低。
Werner Sobek

渲染图可以有效地烘托空间氛围,但其并不能隐藏设计或功能的缺陷。相反,它将缺陷暴露无遗。只有当方案设计非常令人信服时,渲染效果才能抓住人心。一张简单的渲染图无法欺骗业内评委。

Kaspar Kraemer

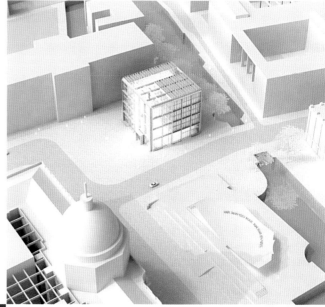

左页上图:奥地利维也纳洲际酒店、滑冰俱乐部、音乐厅建筑竞赛第一阶段,Coop Himmelb(l)au 建筑事务所的方案设计

左页下图:德国柏林 50 Hertz 公司总部大楼建筑竞赛,alexa zahn 建筑事务所的方案设计

上图:德国柏林 50 Hertz 公司总部大楼建筑竞赛,LOVE architecture and urbanism 建筑事务所的方案设计

中左图:德国慕尼黑 BMW FIZ Future 建筑竞赛第一阶段,Clive Wilkinson 建筑事务所的方案设计

中右图:德国柏林蒂森克房伯房地产有限公司办公楼建筑竞赛,Kaspar Kraemer 建筑事务所的方案设计

下图:奥地利维也纳洲际酒店、滑冰俱乐部、音乐厅建筑竞赛第二阶段,querkraft 建筑事务所的方案设计

渲染图讲述了建筑的故事，它也许是最具感染力和吸引力的视觉媒介。任何人都可以把渲染图及其内容联系起来，就如同沉浸在载满过往记忆的绘画或照片中。建筑图纸需要一定程度的"转译"。因此，渲染图是了解和参与建筑体验最重要的视觉媒介。

Daniel Sundlin, Partner at BIG NYC

相比较模型，渲染图非常关键。模型缺少细节刻画，而渲染图可以清晰、详细地表达方案细节，从而弥补模型的缺陷。渲染图也可以帮助表达方案设计的其他部分。需要注意的是，渲染图应该是方案设计的真实表达。

Ben van Berkel

渲染图是一种工具，它可以将非建筑师和评委拉到同一水平线上；它常常诱导并模糊一些设计概念中清晰的部分。借助渲染图营造方案氛围远远超过了方案本身的初始表达。

Louis Becker

渲染图既不能被妖魔化，也不能成为方案表达的绊脚石。需要明确的是：方案设计在没有渲染图的情况下依然可以成立。如今，渲染图作为一种视觉表达媒介，其作用不可或缺，如同建筑的设计与施工，渲染图是时代进步的体现。如同一本杂志，我们既不能逃避它，也不能拒绝或干扰它所具有的信息流通能力（比如将其排除在社会体系之外）。

Thomas Hoffmann-Kuhnt, Publisher, Wettbewerbe Aktuell

如今，渲染图的重要性已显著提高。具有专业背景的评委也许不再依赖渲染图来作出最终决策，但也会或多或少地受其影响；非专业的评委需要借助图像化的表现方式帮助自己以与专家相同的标准评价参赛作品。有一个问题是，渲染图已愈发脱离建筑本身的真实状态，充其量仅仅是一种烘托方案氛围、丰富画面形象的艺术工具，其运用数码照片捕捉建筑的真实状态，但结果却总与最初意图相差甚远。

Markus Allmann

渲染图就像一个无法摆脱的恶魔，如此魅惑，对于非建筑师而言比平面图更易阅读。不幸的是，如今，渲染图变得不可代替并具有误导性。它把建筑浓缩成一个无法完整表现建筑结构复杂性的图像。然而，在这方面，模型更令人信服，它对建筑结构的复杂性毫无欺瞒。因此，评委应该基于模型作出最终决策，而非渲染图。

Frank Barkow 和 Regine Leibinger

左页：奥地利维也纳洲际酒店、滑冰俱乐部、音乐厅建筑竞赛第一阶段，all Design 建筑事务所的方案设计
上图：俄罗斯莫斯科理工博物馆建筑竞赛，FARSHID MOUSSAVI 建筑事务所的方案设计

轴测图，平面图解，图形符号

轴测图一般用于方案设计中的三维空间表达：

- 功能分区；
- 空间动线；
- 交通组织；
- 技术基础设施。

功能分区（竞赛任务所设定的）在轴测图中清晰、易读。图中的不同颜色代表不同的功能分区，颜色与表格的对应使评委可以迅速、容易地解读方案设计。

方案设计的交通图解（例如，员工与访客的交通动线，生产流程和货运路线）在简单的空间动线图上一目了然。

类似的图解元素同样可用于表达方案设计的其他部分。例如，图解交通系统或技术服务说明有助于阐述一个城市规划方案中的相关内容。

可供选择的图形范围很广。其目的只有一个，即每个说明、每个标志清晰、易懂。

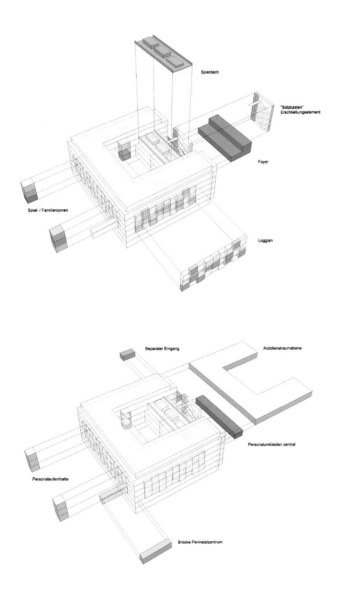
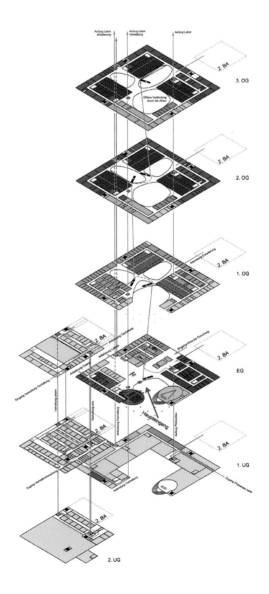

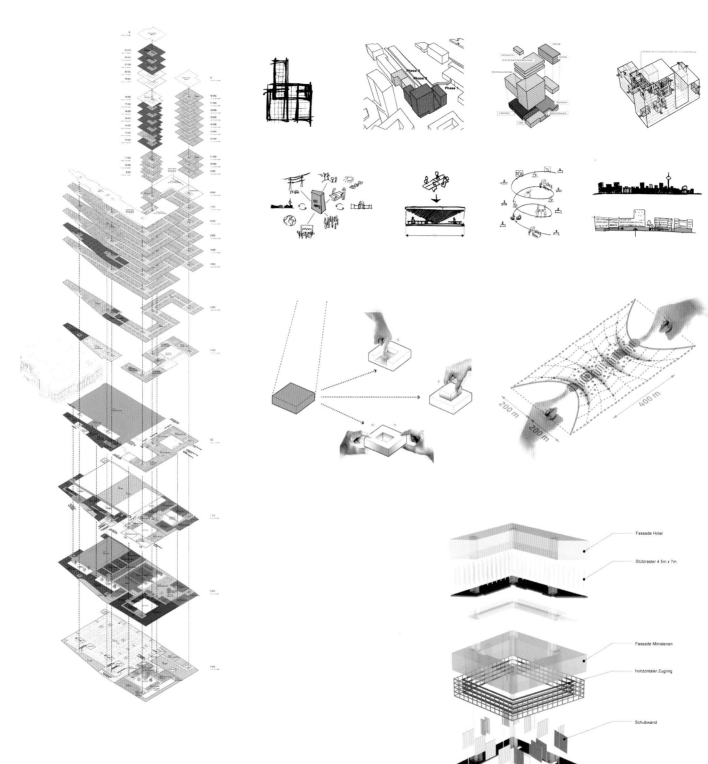

左页左图：德国埃森大学附属医院儿科门诊及核医学研究大楼建筑竞赛第二阶段，Heinle、Wischer 及其合伙人建筑事务所的方案设计

左页右图：德国神经退行性疾病中心建筑竞赛，hammeskrause 建筑事务所的方案设计

左图：奥地利维也纳洲际酒店、滑冰俱乐部、音乐厅建筑竞赛二阶段入围赛，Atelier d'architecture Chaix & Morel 建筑事务所的方案设计

右上图：德国柏林 50 Hertz 公司总部大楼建筑竞赛第二阶段，Henning Larsen 建筑事务所的方案设计

中图：德国慕尼黑 BMW FIZ Future 建筑竞赛第二阶段，Henn 建筑事务所的方案设计

右下图：德国斯图加特卡尔广场邻近地区建筑竞赛，Rafael Viñoly 建筑事务所的方案设计

总平面图

在整体表达中，总平面图是项目介绍及概述的主要媒介，是除模型与三维绘图之外最受评委与业主关注的部分，经常被放于演示文稿的首页。

根据项目的主题，总平面图展现了建筑的屋顶平面和景观元素以及城市尺度的相关信息，例如，景观元素、城市脉络和交通系统的整合。在城市规划竞赛中，总平面图同样包括规划理念的解释说明。

通常情况下，总平面图采用 1:500 的比例，大型项目可以采用 1:1000 的比例。更大规模的城市尺度，采用 1:5000 的比例效果更佳。

总平面图的表达方式多种多样，包括色彩、阴影、线、面、拼贴，等等。

总平面图包括以下内容：

- 建筑的屋顶平面；
- 建筑楼层数量和建筑海拔高度；
- 高程点和轮廓线（若有需要）；
- 交通区域；
- 包括树木和其他重要元素的景观规划。

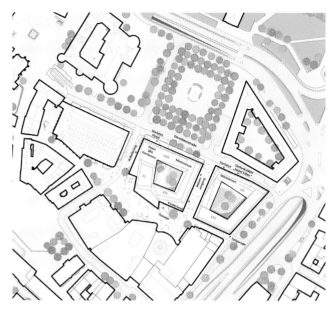
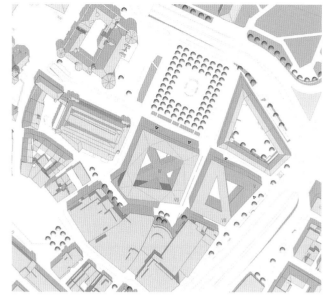

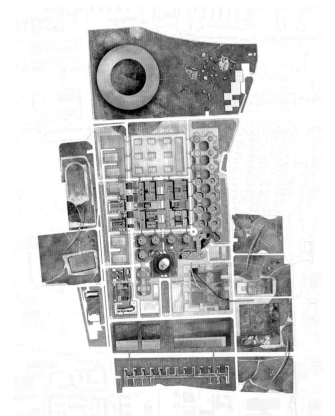

左页：德国柏林蒂森克虏伯房地产有限公司办公楼建筑竞赛，JSWD 建筑事务所的方案设计

上图：德国斯图加特卡尔广场邻近地区建筑竞赛，Behnisch 建筑事务所和 Blocher 及其合伙人建筑事务所的方案设计总平面图（自左向右）

中图：德国慕尼黑 BMW FIZ Future 建筑竞赛第一阶段，KAAN 建筑事务所的方案设计

右下图：德国柏林席勒公园房地产开发竞赛，Bruno Fioretti Marquez 建筑事务所、HAAS 建筑事务所以及 blauraum 建筑事务所的方案设计总平面图（自左向右）

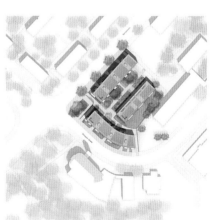
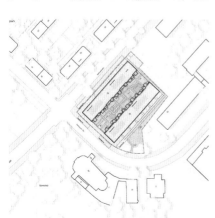

平面图

平面图是一种抽象的二维绘图，展示建筑特定楼层的布局安排。如同标准施工图，平面图是水平于楼层地面上方 1 m 处的剖切面图。尽管如此，平面图也有多种表达方式，这取决于比例和细部设计深度。平面图通常包含线条、色彩、阴影等。作为竞赛成果提交的标准平面图的比例为 1:200，这需要在前期概念方案阶段加以确定。当多数楼层有重复的平面布局时，平面图可以采用较小的比例（1:500），或更简单地，直接采用典型的标准层平面图。这些做法同样适用于双阶段竞赛第一阶段；在此类竞赛中，平面图的细节通常较少。

除建筑结构外，楼层平面图应该包含以下信息，以传达设计理念：

- 交通空间（楼梯、电梯）；
- 可达性（入口、车道等）；
- 标识（房间号、功能分区）；
- 首层周边的室外设施；
- 立面的海拔高度与相对高度；
- 剖面图剖切位置的示意；
- 新建建筑、既有建筑、拆除建筑的图示。

具体的平面设计同样需要一些 1:50 或 1:20 的平面详图（例如，办公室、实验室或公寓中的标准化家具）。通常情况下，它们与 1:100 的图纸的区别主要在于增加了家具或设备。

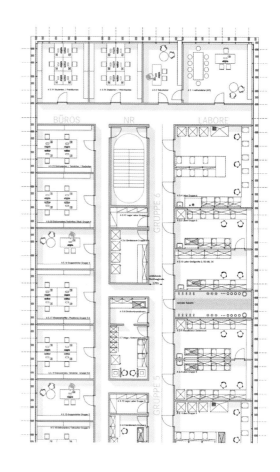
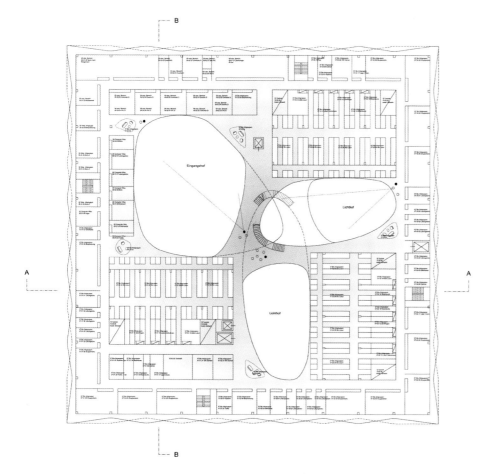

左页：俄罗斯莫斯科理工博物馆建筑竞赛，MASSIMILIANO FUKSAS 建筑事务所的方案设计

左上图：德国柏林 50 Hertz 公司总部大楼建筑竞赛，LOVE architecture and urbanism 建筑事务所的方案设计

右上图：德国尤里西生物研究所生物研究园建筑竞赛，建筑师 Hascher Jehle 的方案设计

下图：德国神经退行性疾病中心建筑竞赛，hammeskrause 建筑事务所的方案设计

立面图与剖面图

平面图、立面图与剖面图，作为三种最基本、最常见的建筑表达方式，能够清晰地展示方案设计的质量，也是评价方案设计的重要参考标准。

立面图展示了建筑的"脸面"和一个非折叠的立面（几个立面连为一体），即建筑的整体形态。同时，立面图也可以完整地展示建筑质量以及建筑与场地及周边环境的相互关系。立面图与剖面图应该至少提供周边环境及建筑的信息，例如，立面开洞比例以及材料和色彩细节。

剖面图，即纵向剖切图，显露了建筑内部空间，包括建筑结构和平面布局的相关信息。通常情况下，方案设计至少应该包含横剖面图和纵剖面图。剖面视图可以将横剖面图和纵剖面图中的各种元素巧妙、完整地结合在一起。

此外，立面图和剖面图还可以展示建筑平面对海拔高度的精准测量。立面图和剖面图的比例通常与平面图的比例保持一致。

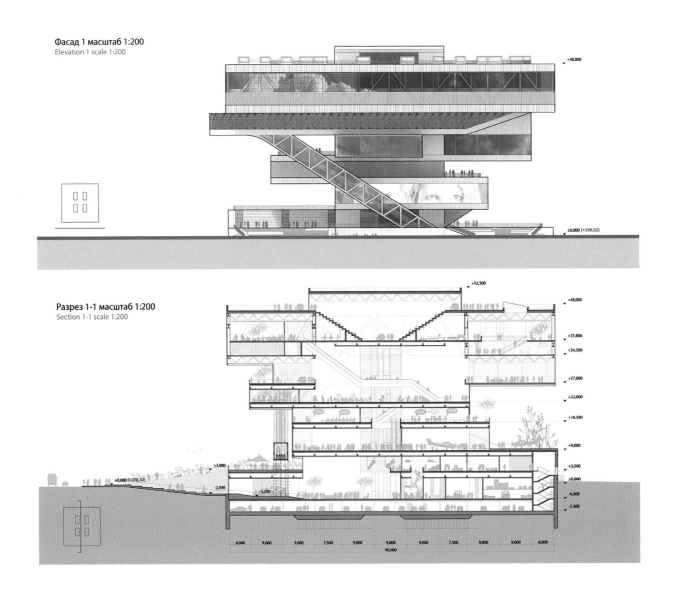

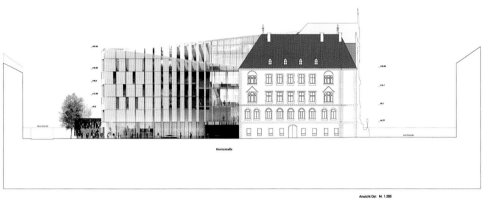
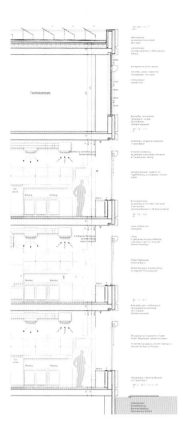
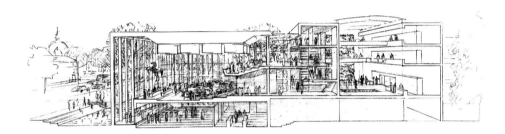
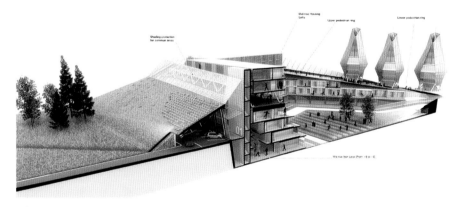
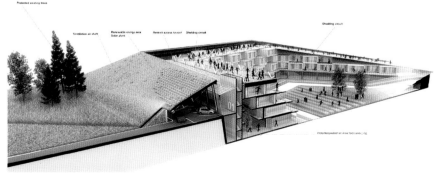

左页：俄罗斯莫斯科理工博物馆建筑竞赛，MECANOO International 建筑事务所的方案设计
左上图：德国威斯巴登法学院建筑竞赛第二阶段，3XN A/S 建筑事务所的方案设计
右上图：德国尤里西生物研究所生物研究园建筑竞赛，Atelier 30 建筑事务所的方案设计
中图：德国汉诺威下萨克森州议会大厦建筑竞赛，mm architekten 建筑事务所的方案设计
右图：加拿大马尼托巴大学校园及周边地区建筑竞赛第二阶段，IAD Independent Architectural Diplomacy 建筑事务所的方案设计

可持续发展与工程技术

大多数竞赛专注于城市规划、建筑设计或景观设计,但在可持续发展与工程技术方面并没有明确的规定。可持续发展与工程技术在参赛作品评审过程中同样也扮演从属的角色。

可持续发展与工程技术的相关内容经常被缩减至几个图解或几张图片,且与能量论述相结合。图解或图片极度受限的交流价值与相对受限的工程咨询投入密切相关。

尽管如此,看起来非常表面化的图解或图片及其效果却不应被低估,它们在方案表达中具有特殊的作用。在评委看来,它们体现了方案设计的基本理念和具有可持续性的综合规划方法。

专业顾问的引入在常规投标竞赛中起到更加重要的作用,他们对建筑评价和检测作出权威性的文字说明和图解。

构架:

• 结构体系示意图;
• 主结构系统图解(轴测图)。

能量概念:

• 水电系统在水平与竖直方向上的布局,管径容量评估;
• 在城市结构中,日照方向和风向,周边建筑和土地关系的相互整合;
• 暖通空调系统和能量供给系统图解(电力、应急供电、供热、冷却);
• 简明的能量流动图解(桑基图)显示相关的季节性能量流;
• 在细部剖立面图中,结构和能量的相互整合(例如,冬、夏两季温度和阳光控制)。

为了贯彻"可持续发展"的原则,在适当的阶段应考虑以下标准:

• 环境可持续性——可持续性城市规划和景观设计,可持续性建筑尺度和技术元素(例如,立面细节和技术概念);
• 社会可持续性——特殊措施的介绍;
• 经济可持续性——为建筑业务提供成本-效益解决方案。

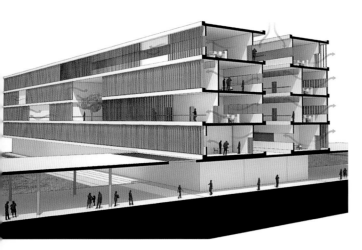
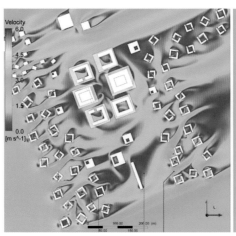

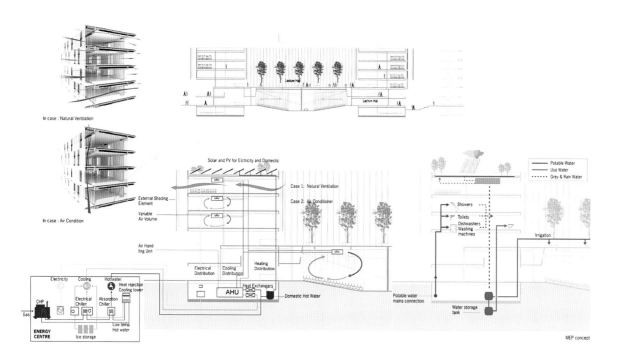

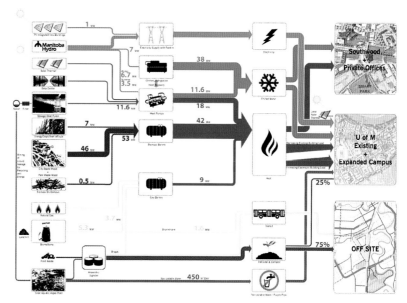

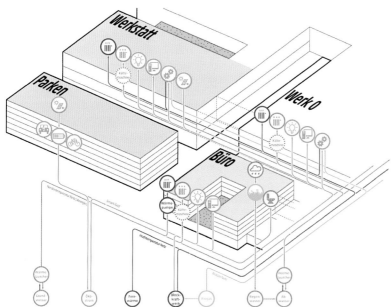

左页左图：越南 - 德国大学校园建筑竞赛第一阶段，KSP Jürgen Engel 建筑事务所的方案设计

左页右图：越南 - 德国大学校园建筑竞赛，Henn 建筑事务所的方案设计

上图：加拿大马尼托巴大学校园及周边地区竞赛第二阶段，Perkins+Will 建筑事务所的方案设计

中图：越南 - 德国大学校园建筑竞赛第二阶段，KSP Jürgen Engel 建筑事务所的方案设计

下图：德国慕尼黑 BMW FIZ Future 建筑竞赛第二阶段，Henn 建筑事务所的方案设计

个人竞赛方案演示

在建筑师和规划师的职业生涯中,个人方案演示十分普遍。然而,竞赛评审很少设置演示环节,这是一个例外。有调查显示,即使在匿名竞赛中,个人方案演示对于赢得竞赛与否也至关重要——一般在竞赛结束之后、签订合同之前。以下是赢得竞赛和签订合同的中间环节,个人方案演示在此阶段不可或缺甚至非常关键:

• 组委会会面,与竞赛赞助方协商合同事宜;
• 记者招待会;
• 市民团体会议。

在非匿名竞赛中,无论成果演示还是过程演示,设计团队都应该被详细介绍。在合同协商期间,出席的设计团队关键成员十分重要,通常有相应的遴选标准。个人方案演示的关键在于激起观众的兴趣,获得其对于设计理念与想法的认可和赞同。

演示

个人方案演示通常采用口头表达和视觉呈现的方式,二者相互穿插。借助于一系列表达媒介(无需过多的技术操作),演示语言通俗、易懂。当然,这同时也展示了演示者本人对于方案设计的自信。众多视觉呈现元素在本质上具有服务性。"少即是多"的理念体现在幻灯片的数量和每页幻灯片的文字说明上。观众无需阅读演示文件就可被自然地带入方案思路中。

根据以往经验,演示者应尽量避免过多地使用演示文稿程序中的特效功能。方案演示应该言简意赅,以"观众理解"为目的。若要使方案演示易于理解,可在幻灯片中增加一些单一的图案或流动的箭头来模拟空间交通动线,等等。

个人形象

在有关个人方案演示的实用技巧和成功案例中,演示者给观众留下的第一印象尤为重要。针对每个特别情况的充分准备和对文稿程序的深入研究是演示成功的前提——无论对于初次参赛的"菜鸟"还是多次参赛的老练的"超级明星"。与演示相关的诸多问题没有普适的解决方案,但当演示者未能解决相应问题时,观众一定会注意到以下问题:

• 演示的观众是谁,是业内同行还是外行?观众是小圈子的听众,还是更大范围的评审委员会,抑或大型开放礼堂中的听众?观众的构成影响着设计团队参与演示的成员、演示者的衣着及演讲质量。除此之外,问候语、肢体语言、语速、音量、汇报位置以及观众的座位排布都应提前予以考虑。脱稿演示的效果通常胜过照本宣科,但有时阅读一份准备充分的文稿比自由谈话更令人信服。

• 演示者想传达什么？方案设计的核心部分应该成为演示过程中的线索。演示的重点内容应该精心选择并以易于理解的语言表达出来。

• 演示者如何让评审委员会相信其所提供的方案设计是众多参赛设计中的最佳选择？演示者如何让业主相信其是最好的工作伙伴并能为业主带来最具特色的"头脑风暴"？相比较方案设计理念或实施过程，演示者发自内心的自信和热情更加重要，甚至堪称获胜的关键。这说明，演示者的个人魅力高于一切。在个人方案演示中，一位年轻的建筑师不应该把自己表现得像一个当红明星或高高在上的哲学家，而应以技术为专长，在永远更新且学无止境的建筑技术面前保持谦逊的态度。此外，每个项目的成功都得益于团队合作。几个人在演示过程中的协调互动非常重要，指定的项目负责人应该参与演示。如果没有足够的时间留给其他成员，那么他们至少都要被一一介绍。

学界有很多关于个人方案演示成功案例的介绍与指南，参赛者可以在日常学习、工作中搜罗一些，以供参考。

设备与工具

在个人方案演示之前，以下几点需要确认无误：

• 使用哪些文稿及图画程序或软件，例如 Acrobat Reader、Microsoft PowerPoint 或 Prezi？

• 如果使用个人电脑，需要用什么适配器连接设备？如果设备不兼容，那么有什么应急预案？

• 多大容量且多少数量的幻灯片是被允许的？还需要其他控制监视器吗？

• 音频或视频是否需要被集成控制？如果需要，那么现场需要什么技术设备？

• 演示文稿（设计文本及说明）是否需要印制成册并现场发放？

上图：德国 GMP 建筑事务所的建筑师 Nikolaus Goetze 在德国慕尼黑 BMW FIZ Future 建筑竞赛第二阶段中做个人方案演示

 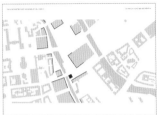  
  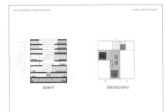 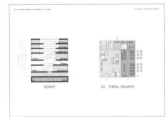
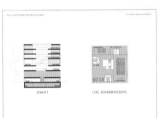 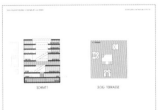 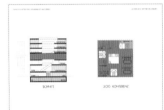 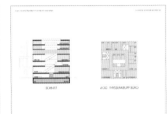
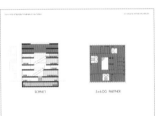   
 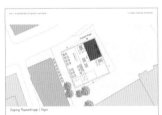  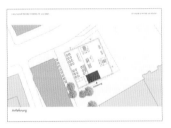

 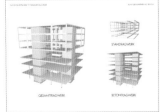 

Schweger 及其合伙人建筑事务所（位于德国汉堡）在德国柏林蒂森克虏伯房地产有限公司办公楼建筑竞赛参赛者学术研讨会上就个人方案所做的幻灯片演示

对于评委而言，个人方案演示的重中之重是什么？对于演示者而言，其又是什么？

这个问题非常重要。事实上，没有个人方案演示的竞赛根本不应该举办。

Werner Sobek

面对面是个人方案演示的最佳形式，演示者可以在演示过程中收集观众所反馈的信息。演示者、评委及业主彼此都能发现对方正在做什么，如果设计理念与想法是兼容的，那么就产生了所谓"化学反应"。对话的效果远远胜过揣测。

Frank Barkow 和 Regine Leibinger

将组织规程与个人演示相结合很可能得出一些相互矛盾的结论或评价。与业主进行建设性对话可能使匿名竞赛丧失"匿名"的意义，进而失去客观性。

Markus Allmann

个人方案展示十分重要：一方面是阐述设计；另一方面，业主并非简单地选择一个方案，而是慎重地选择一个未来长期合作的伙伴。

Jakob Dunkl

个人方案演示对所有参与者都有好处：业主与建筑师可以测试其对待定的方案进程所面临的机会与风险是否达成共识，这也是竞赛过程中开放性讨论的环节之一。

Thomas Willemeit

在进行个人方案演示时，有太多可以沟通、交流的东西。竞赛存在的理由让个人成果得以更加清晰的展示。回答评委的提问和与评委互动为建筑师在赛后继续深化设计提供了指引。

Louis Becker

个人方案演示为业主与建筑师提供了一个当面对话的机会。在个人方案演示的下个阶段，业主可以向建筑师咨询一些关于项目设计理念及操作方法的具体问题；建筑师在竞赛最后阶段将业主所反馈的信息融入项目的深化设计中，使其更符合业主的需求。

Ben van Berkel

上图：Schultes Frank 建筑事务所的 Axel Schultes 教授在德国柏林蒂森克虏伯房地产有限公司办公楼建筑竞赛参赛者学术研讨会上做个人方案演示

左页：West 8 都市景观事务所的设计师 Christoph Elsässer 在德国慕尼黑 BMW FIZ Future 建筑竞赛第三次参赛者学术研讨会上做个人方案演示

竞赛成果制作、提交及其他细节

竞赛结束时，除了展示图版，成果提交包括内容和形式两方面：作者声明、项目数据图表、相关的数字文档、设计文本及说明，等等。

设计文本及说明在竞赛中的被关注度往往取决于初审员和评委。冗长的文本及说明很可能被忽视，因此，其不必过长，应该简短、易读、全面，否则被"视而不见"的风险就会大大增加。设计文本及说明具有诠释方案设计理念、技术应用及实施过程的作用，其常常作为项目后续动态发展的基础，因此应该有适当的叙述标准和长度、深度限制。在跨学科竞赛中，设计文本及说明具有更多的意义，方案设计的核心理念及技术水平等关键部分均应在设计文本及说明中有所体现甚至被重点描述。

在竞赛第一轮中，解释性文字扮演从属的角色。它们只有在初审出现问题（这种情况十分罕见）时才会被研读。文字说明只有在作品精美且清晰展现的情况下才会被阅读，并且是在作品入围之时。好的文字说明对于设计灵感追溯和设计思维展示所起到的作用十分有限。

Kaspar Kraemer

只有入围作品的文字说明才会被更多地关注。

Markus Allmann

设计本身是独立的，不需要大段的文字说明，从平面图中就可以看出一名优秀建筑师的鲜明特征，而不是他／她写的一行行文字说明。

Thomas Hoffmann-Kuhnt, Publisher, Wettbewerbe Aktuell

评委经常对最后几个参赛作品不太关注。若要在众多参赛作品中脱颖而出，文字说明必须言简意赅。

Jakob Dunkl

运输与物流

在竞赛中，图纸和模型的运输是必要的。图纸和模型在多次转运路线（从初审地到评委会所在地再到展览大厅等）中可能被损坏，坚固且可重复使用的打包方式可有效地减小其被损坏的可能性。相比较图纸和数据库，模型尤其容易因不当的打包方式而被损坏。通常情况下，可以采用以下方式避免其被损坏：

- 首先用塑料将图纸卷起包上，保护好纸张的边缘，然后用坚固的卡纸板或塑料图纸筒装好；
- 用标准商用信封寄送技术文档；
- 模型包裹必须坚固并且可以重复使用（使用木板箱或厚板条箱），包裹内只包含模型；
- 模型本身应该有稳定的结构。用胶粘的部分最薄弱，在运输过程中易损坏；
- 模型应固定在包装框架上，以防滑动而损坏；
- 即使在盒子或信封中，CD 和 DVD 作为数据载体也容易开裂。

左页：评审过程中参赛作品的设计文本及说明同绘画图纸并排放置
上图：运输小型模型时所用的木板箱
中图：不恰当的运输方式可能导致模型到达目的地后被损坏
下图：将模型完整地放在坚固的板条箱中，这是运输模型的最佳方式

打印

- 展示图版应该在厚纸上高精度打印（140～200 克 / 平方米）。不建议使用光面纸张，否则容易造成反光。
- 所有技术文档（包括为初审准备的图版备份和缩小版复印件）均应在标准纸张上打印。
- 建议把待提交的图纸放在图纸桶里，不要裱在板子上。图纸裱板容易弯折，在评审过程中不易悬挂，而且运输成本高。因此，只有在评审过程有特殊需求时才以此方式提交图纸。
- 在初审之前，所有技术文档不应装订在提交作品上，而应统一复制并归档。

它们不可被压在图纸筒里或粘在包裹外侧。U 盘和存储卡是比较好的选择。

运输应该由建筑师本人或专业、可信的运输人员完成，并保留物流单。国际运输需要海关申报。根据以往经验，在票据中写清"展览样品"或"无商业价值的竞赛模型"可以避免不必要的费用或延误。

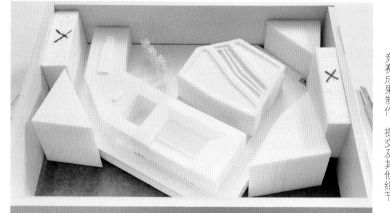

竞赛项目

旧城复兴
—— Lalla Yeddouna 广场

摩洛哥 非斯

2010—2011

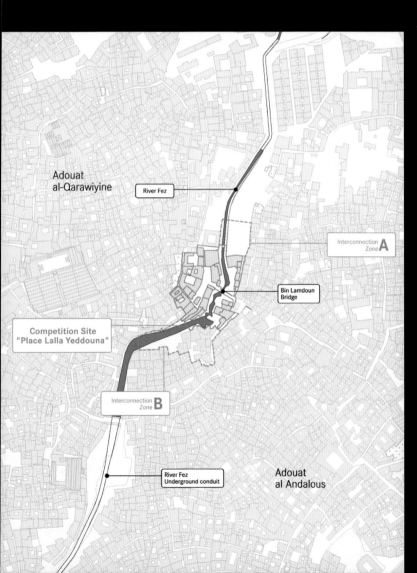

概况
业主：摩洛哥王国
项目规模：基地面积约 7400 平方米
类型：开放型双阶段设计竞赛（开放申请程序）
参赛者：第一阶段 176 组，第二阶段 8 组
竞赛预算：440 000 美元（第二阶段奖金 120 000 美元，第二阶段每组参赛者的费用 40 000 美元）

评委
建筑评委：Prof. Dr. Marc Angélil, Architect, Zurich; Meisa Batayneh Maani, Architect, Amman; Stefano Bianca, Architectural Historian, Orzens; Sir David Chipperfield, Architect, London; Omar Farkhani, Architect, Rabat; Prof. Rodolfo Machado, Architect, Boston; Matthias Sauerbruch, Architect, Berlin
专家评委：Abbas El Fassi, Prime Minister of the Kingdom of Morocco; Samuel L. Kaplan, US Ambassador to the Kingdom of Morocco; Bensalem Himmich, Culture Minister of the Kingdom of Morocco; Anis Birou, State Secretary for Handicraft of the Kingdom of Morocco; Mohammed Rharrabi, Governor of the Fez-Boulemane Region; Abdelhamid Chabat, Mayor of Fez

"在未来,Lalla Yeddouna 将成为一个供北非阿拉伯居民和游客使用的城市中心,充满活力并且功能齐备;该地是工艺发展的重要触媒,为年轻人和年长者提供活动空间。"

——引自《竞赛摘要》

工匠和非斯阿拉伯人聚集地项目
该项目由世纪挑战账户集团(MCC)组织,受摩洛哥政府和美国政府的资助,旨在复兴对古镇旅游业和当地经济尤为重要的五个地点(Lalla Yeddouna Square、Chemmayine、Sbitriyine、Staouniyine、El Barka Fonduks),它们均位于非斯旧城(北非阿拉伯人聚集地)。经济发展有助于改变当地居民的贫困状态,为其营造良好的生活和工作环境,使这里成为"可持续生活与工作之所"。同时,该项目的建设也间接地保护了当时丰富且历史悠久的文化与建筑遗址。

Lalla Yeddouna 广场
7400 平方米的竞赛区域坐落于非斯阿拉伯人聚集地。非斯是摩洛哥第三大城市,也是该国四个古帝国城市中最古老的一个。作为一个典型的东方城市,非斯于 1981 年被列入联合国教科文组织世界遗产地名录。马赛克式小型沙土颜色的房子和迷宫般的狭窄街道是传统、古老的生活方式的象征,加之墙体组织,这里成为活生生的博物馆。围绕 Lalla Yeddouna 广场的街坊距离 8 世纪的安达鲁西亚之角很近,那里是该区域最古老的部分,因其传统手工艺、制革、黄铜和铜铁匠铺而闻名。该区域工作与生活的紧密结合恰如其分地彰显出项目特征。

竞赛任务
运用"城市-建筑"策略,加强街道附近广场和滨河地段"社会-经济"结构的示范意义。如果不考虑地域和人口密度的具体特征,项目应构建一个新的模型,将有价值的物质空间及实体予以恢复并重新整合,在异常狭窄的空间体系内扩建并维护全新的小型建筑,并创造持久的生命力。

第一阶段参赛者

第二阶段入围者

1. hanse unit, Hamburg (D)
2. mossessian & partners architecture, London (GB)
3. Studio Ferretti-Marcelloni, Rome (I)
4. Moxon Architects, London (GB)
5. Studio Giorgio Ciarallo, Rho (I)
6. kolb hader architekten, Vienna (A)
7. Hashim Sarkis Studios, Beirut (RL)
8. Arquivio Architectura, Madrid (E)

其他参赛者

9 mari–as arquitectos asociados, Seville (E) **10** Schneider + Sendelbach Architekten, Braunschweig (D) **11** N+B architectes, Elodie Nourrigat & Jacques Brion, Montpellier (F) **12** Boyarsky Murphy Architects, London (GB) **13** Dr. Maged Aboul-Ela & Architect Eman Hatem, Cairo (ET) **14** Abdullah Kocamaz, Istanbul (TR) **15** FGA Arquitectos, Mexico City (MEX) **16** HOK International, Hong Kong (HK) **17** Kraaijvanger · Urbis, Rotterdam (NL) **18** Kuehn Malvezzi, Berlin (D) **19** maxwan, Rotterdam (NL) **20** JZMK PARTNERS, Irvine (USA) **21** pbr Planungsbüro Rohling Architekten und Ingenieure, Osnabrück (D) **22** Bularch, Sofia (BG) **23** Olekov Architects, Sofia (BG) **24** andruetto deri architetti associati, Pisa (I) **25** Larkin Architect, Toronto (CDN) **26** ORTIZ MONASTERIO + ASOCIADO, Mexico City (MEX) **27** PurserLee, Dallas (USA) **28** IXII, Kobe (J) **29** Never Ending Architecture, N.E.A., Jerusalem (IL) **30** scapelab, Ljubljana (SLO) **31** Taller 301, Bogota (CO) **32** Geurst & Schulze Architecten, The Hague (NL) **33** cabinet Seqqat Nabila, Meknès (MA) **34** MM26, Padua (I) **35** Reardon Smith Architects, London (GB)

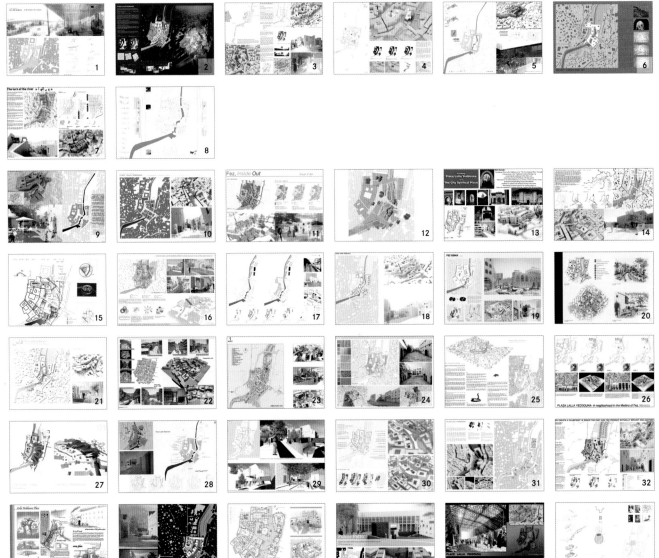

36 Museum The Garage, Rotterdam (NL) **37** SARMA & NORDE Architects, Riga (LV) **38** LOADINGDOCK5 ARCHITECTURE, Brooklyn (USA) **39** KOLLIAS GEORGE, Heraklion (GR) **40** R&Sie(n) Architects, Paris (F) **41** Vaknine Architects & Town Planners, Jerusalem (IL) **42** Groupe3Architectes, Rabat (MA) **43** ARJM Architecture, Brussels (B) **44** ATELIER 8000, Prague (CZ) **45** Nieto Sobejano Arquitectos, Madrid (E) **46** Sergio Pascolo Architects, Venice (I) **47** Haptic, London (GB) **48** Filipe Oliveira Dias Arquitecto, Porto (P) **49** Mezger & Schleicher, Stuttgart (D) **50** dennis ulm architekt, Munich (D) **51** KOKO architects, Tallinn (EST) **52** arkiteyp, Istanbul (TR) **53** Peter Tagiuri, Architects, Cambridge (GB) **54** TENDANCES, Tunis (TN) **55** DURU & PARTNERS, Montpellier (F) **56** Agence Fikri Benabdallah Architecte, Rabat (MA), Federico Wulff & Melina Guirnaldos Arquitectos, Madrid (E), ARCHITEXTURS and Rachid Andaloussi Ben Brahim Office, Rabat (MA) **57** Architectural Bureau Zotov & Co., Kiev (UA) **58** Ritzen Architecten, Maastricht (NL) **59** Berger + Parkkinen Architekten, Vienna (A) **60** N.E.E.D., New York (USA) **61** H3T Architekti, Prague (CZ) **62** Reaction Architecture, Tunis (TN) **63** ARCHLAB, Monopoli (I) **64** Davide Rapp, Milan (I) **65** LAAP Landscape + Architecture, Arquitectura + Paisaje, Mexico City (MEX) **66** TSARA Architectes, Clichy (F) **67** Claudio Silvestrin Architects, London (GB) **68** Architekturbüro Albrecht, Munich (D) **69** OMAYAN, Tangier (MA) **70** JCSR Arquitecto, Seville (E) **71** Idrissi Architecture Office, Salé (MA) **72** Durand-Hollis Rupe Architects, San Antonio (USA) **73** Mila/Jakob Tigges, Berlin (D) **74** WindStone International, Berlin (D) **75** Atelier Moto Katono, Tokyo (J) **76** Enlace Arquitectura, Caracas (YV) **77** SMC Management Contractors, Nicola Fazio, KE architekten and Francesco Minnitti, Winterthur (CH) **78** Peter Kellow Architecture, Plymouth (GB) **79** ma.lo architectural office, Innsbruck (A) **80** S.A.E.C., Naples (I)

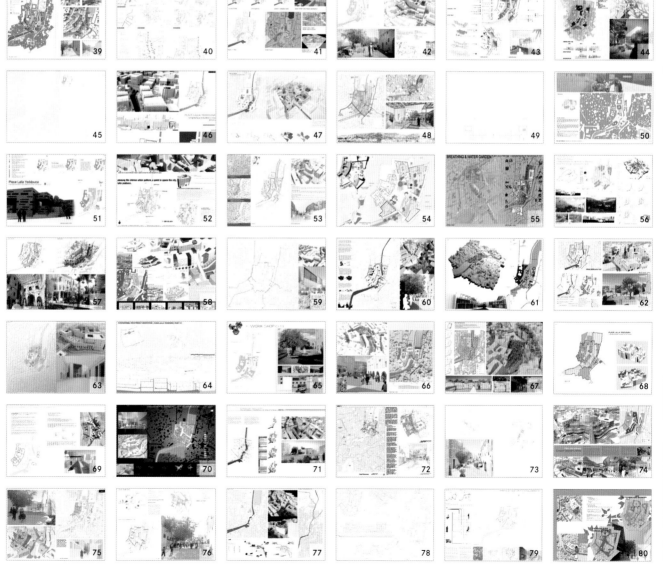

第一阶段其他参与者

81 Dinmez Insaat Sanayi, Izmir (TR) **82** SHArchs, Cincinnati (USA) **83** Serero Architectes Urbanistes, Paris (F) **84** paula santos arquitectura, Porto (P) **85** Valerio Morabito, Reggio Calabria (I) **86** POLY RYTHMIC ARCHITECTURE, Bordeaux (F) **87** EWA, London (GB) **88** Lino Bianco and Associates, Hamrun (M) **89** Carmelo Bagalà Architects, Milan (I) **90** MEMA Arquitectos, Bogota (CO) **91** moh architects, Vienna (A) **92** Adriano De Gioannis, Rome (I) **93** HCP Architecture & Engineering, Malaga (E) **94** Architecture Project, Valletta (M) **95** Consolidated Consultants - Jafar Tukan Architects with TURATH, Amman (JOR) **96** Eleena Jamil Architect, Ampang (MAL) **97** WAP Architects, Sheffield (GB) **98** NED University of Engineering & Technology, Karachi (PK) **99** HANDIS, Casablanca (MA) **100** arb east architects, Hanoi (VN) **101** diaspora, Coimbra (P) **102** DP Architects, Singapore (SGP) **103** Limin Hee, Singapore (SGP) **104** CKM, Singapore (SGP) **105** CAMPO aud, Rio de Janeiro (BR) **106** Armstrong + Cohen Architecture, Gainesville (USA) **107** AiB estudi d'arquitectes, Barcelona (E) **108** Bassen Sauletshileri, Almaty (KZ) **109** manzl ritsch sandner architekten, Innsbruck (A) **110** LOVE architecture and urbanism, Graz (A) **111** Studio K, Naples (I) **112** Fundacion CEPA, Buenos Aires (RA) **113** Lostmodern, Paris (F) **114** K.N.Z design architecture & space, Amman (JOR) **115** HOSHINO ARCHITECTS, Tokyo (J) **116** guillaume girod architecte, Grenoble (F) **117** Kengo Kuma & Associates, Tokyo (J) **118** EBA [For Cosmopolitain], Toronto (CDN) **119** ziya necati özkan architectural & engineering office, Nicosia (CY) **120** AdM Arquitectes, Barcelona (E) **121** ATBA atelier baya, Tiznit (MA) **122** FACE2050, Bad Griesbach (D) **123** Independents 499, Buffalo (USA) **124** Klingmann Architects, New York (USA) **125** Uwe Bernd Friedemann, Cologne (D) **126** claudiovilarinho.com architects and designers, Porto (P) **127** elementa architects, Seoul (ROK) **128** Rarcon, Vila Nova de Gaia (P) **129** AAU/A, Paris (F) **130** URBAMED, Paris (F) **131** Pascal Flammer Büro für Architektur, Zurich (CH) **132** Estudio CV_Public Opera, Buenos Aires (RA) **133** Urban Edge Consultants, Rahway (USA) **134** Arriola & Fiol Arquitectes, Barcelona (E) **135** Graeme Massie Architects, Edinburgh (GB) **136** Francesca Mugnai Architetto, Florence (I) **137** JURONG Consultants, Singapore (SGP) **138** Architecture Republic, Dublin (IRL) **139** Ünsal Demir, Istanbul (TR) **140** DRC, Jounieh (RL) **141** Estudi d'Arquitectura Josep Blesa, Valencia (E) **142** CAVstudio, Lisbon (P) **143** DRMS with Marije Tersteege, Amsterdam (NL) **144** Reset architecture, 's-Hertogenbosch (NL) **145** KuKu, Athens (GR) **146** Koschany Zimmer Architekten KZA, Essen (D) **147** Jean-Yves QUAY, Lyon (F) **148** FSA architects, Madrid (E) **149** Yves WOZNIAK Architecte, Marquillies (F)

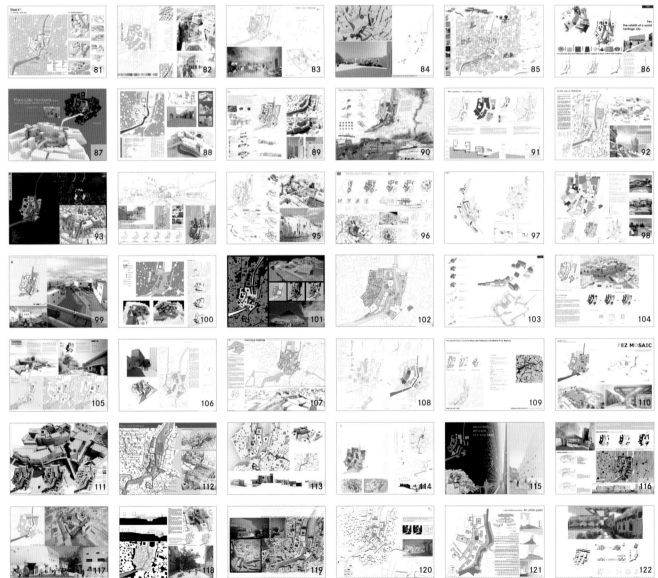

150 PZP ARHITECTURA, Bucharest (RO) **151** MOBA studio, Prague (CZ) **152** SOLIDUM, Medellin (CO) **153** Seo-Kang Architects Office, Seoul (ROK)
154 Arcus Architects, Belfast (GB) **155** Carlos Arroyo Architects, Madrid (E) **156** disart, Barcelona (E) **157** CASALEGANITOS Estudio de Arquitectura, Madrid (E)
158 Architecture & Heritage, Plovdiv (BG) **159** Sudarch, Reggio Calabria (I) **160** Papazian Roy Architecte, Paris (F) **161** Mahmoud Saimeh, Amman (JOR)
162 the fourth dimension, Zarinshahr (IR) **163** SIC Arquitectura y Urbanismo, Madrid (E) **164** Rafał Mroczkowski Architekci, Poznan (PL) **165** Brut Deluxe, Madrid (E) **166** Stephen Collier Architects, Surry Hills (GB) **167** HDAA – Heitor Derbli Arquitetos Associados, Rio de Janeiro (BR) **168** Gaheez Consultants, Peshawar (PK) **169** ACD Studio, St. Petersburg (RUS) **170** Relative Form Architecture Studio, Vancouver (CDN) **171** els architecture, Decatur (USA)
172 icarquitectura, Figueres (E) **173** Hernandez Leon Arquitectos, Madrid (E) **174** Aedas, London (GB) **175** Acharhabi Architecte, Casablanca (MA)
176 architecture & design, Isabelle Gaspard, Paris (F)

第二阶段参赛者

获奖者

1. 一等奖 mossessian & partners architecture, London (GB), with Yassir Khalil Studio, Casablanca (MA)
2. 二等奖 Studio Ferretti-Marcelloni, Rome (I), with Bahia Nouh, Fez (MA)
3. 三等奖 Moxon Architects, London (GB), with Aime Kakon, Casablanca (MA)

第二轮

4. Bureau E.A.S.T., Los Angeles (USA)/Fez (MA), with Atelier3AM, Toronto (CDN), and Hashim Sarkis Studios, Beirut (RL)

第一轮

5. Arquivio Arquitectura, Madrid (E), with Taoufik El Oufir Architectes, Rabat (MA)
6. Studio Giorgio Ciarallo, Rho (I), with Emiliano Bugatti, Istanbul (TR), and Cabinet Hadmi, Safi (MA)
7. hanse unit, Hamburg (D), with Zine El Abidine Lasouini, Fez (MA)
8. kolb hader architekten, Vienna (A), with Kubik Studio, Meknés (MA)

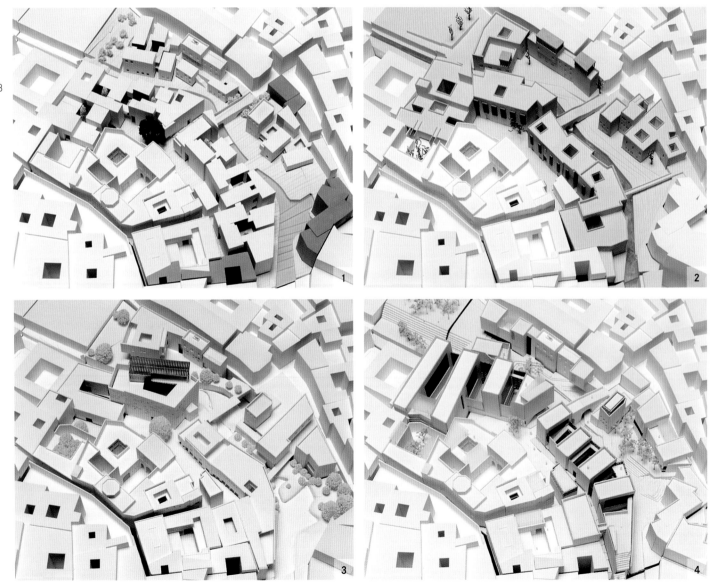

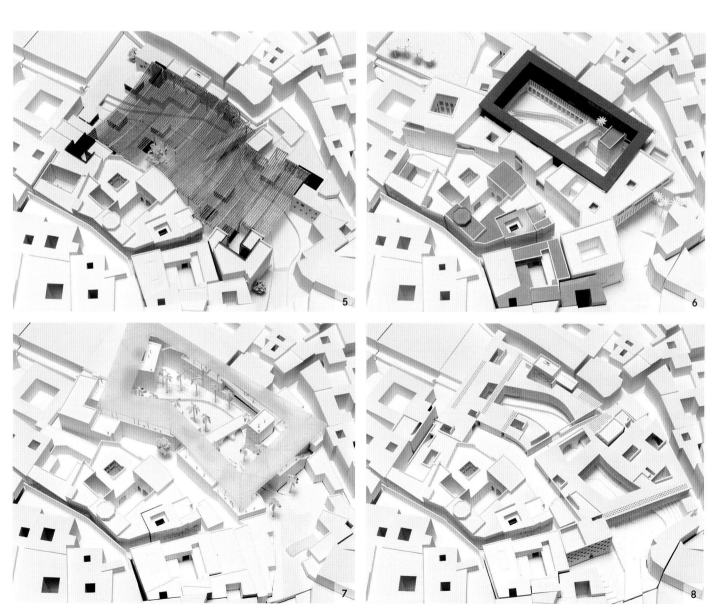

5 6 7 8

竞赛一等奖　获奖事务所

mossessian & partners architecture

London (GB); with Yassir Khalil Studio, Casablanca (MA)

方案设计者： Michel Mossessian, Yassir Khalil
参与者： Jose Marquez, Bulut Cebeci, Selim Bayer, John Veikos, Per Brunkstedt;
Anthony Mopty, Mourad Bellaanaya, Abdelkarim Tounli, Saana El Kahlaoui, Nadia Azia
自由职业者： Maurel Fabrice, Youssef Lahrichi

01 PLACE LALLA YEDDOUNA

010067

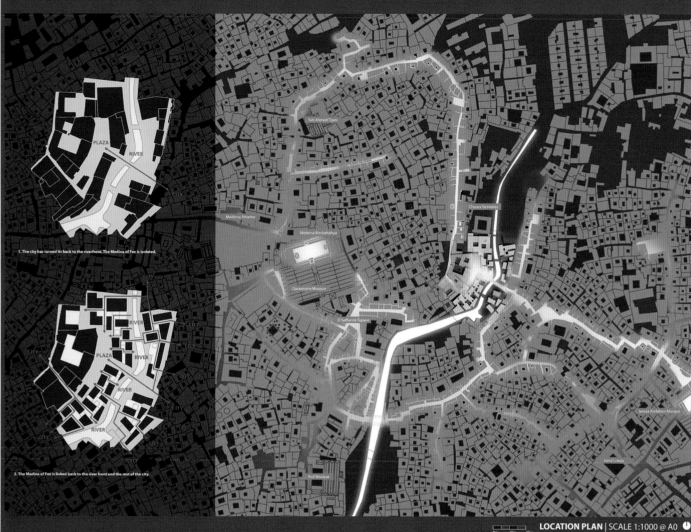

1. The city has turned its back to the riverfront. The Medina of Fez is isolated.

2. The Medina of Fez is linked back to the river front and the rest of the city.

LOCATION PLAN | SCALE 1:1000 @ A0

INTRODUCTION: URBAN POROSITY

The city has turned its back to the riverfront - creating a series of disconnected urban spaces. The two banks of the river are linked via a pedestrian bridge, leaving the riverfront isolated from public activity. Our strategy is to recover the medina urban feel of connecting Place Lalla Yeddouna to the riverfront and across, via internal streets and courtyards, offering nonlinear paths of circulation within the Medina. The Urban porosity offers a hierarchy of continuous spaces within a rich urban pattern of small passages, views and relationship between the Place Lalla Yeddouna and the riverfront.

We have identified three inter-related 'VECTORS' to establish our design Vision for the project:

URBAN ARTICULATION, PROGRAMME, and ENVIRONMENT.

HOW DO WE CREATE A SQUARE THAT MEDIATES BETWEEN PUBLIC AND PRIVATE SPACE, AND IS POROUS WHILST INVITING TO WORK, PLAY AND LIVE?

PLACE LALLA YEDDOUNA

It is important to acknowledge and preserve the historical sense and space of Place Lalla Yeddouna as an 'urban room', where people are encouraged to meet, greet and linger. Whilst the historically private character of the west side of Place Lalla Yeddouna is preserved, the east side takes on a more porous form, where connectivity with the Fez River is provided via alleyways and riads.

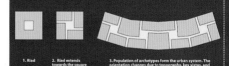

1. Riad
2. Riad extends towards the square and the riverfront, forming an archetype
3. Population of archetypes form the urban system. The orientation changes due to topography, key vistas, and relationship to riverfront. The connected semi open spaces and courts create a new kind of urban space for arts and commerce.

4. Multi level outdoor spaces are created within building mass:

5. Diagrammatic section illustrating levels and uses:

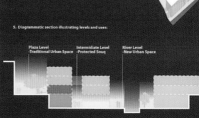

- Artisan Facilities: Workshops, Display and Direct Sales
- Public and Tourist Services: Information Centre, Retail and Exhibition Spaces
- Public and Tourist Services: Restaurants, Cafes and a Hotel
- Municipal Facilities: Management Offices, a Post Office and Fire Brigade

URBAN ARTICULATION: SCULPTING THE VOID

Urban Pattern:
Our intent is to 'prime' space between buildings rather than propose building objects. 'Sculpting the Void' relates to the aspiration that buildings offer a continuous framework for people to work, meet and connect. Buildings are background to the activities they offer.

Private vs Public spaces
Traditional Islamic cities are built around a network of spaces set in a hierarchy from private inner space to communal public space. There is no distinct separation of public and private: public space pervades each private enclave in the form of the 'majlis', internal courtyards and winding residential streets. There is also a symbolic relationship between public and private: each smaller space is a microcosm of larger spaces, the 'majlis' (as the centre of the home) mirrors the courtyard (as the centre of a building) which in turn mirrors the public space (as the centre of the community). This cellular network of spaces is the foundation for the unity of a neighbourhood. In this respect, it is helpful to think of the development as a single building with Place Lalla Yeddouna, as an 'urban room' at its heart.

HOW CAN WE UNITE A FRAGMENTED URBAN SPACE WHILST PRESERVING ITS HISTORY, IDENTITY AND CHARACTER?

PROGRAMME: LANGUAGE AND MEANING

Moroccan vernacular architecture is based on regular but adaptable typologies. We have introduced variation and accents into these traditional forms to give a sense of individuality and place and accommodate the dense arrangement of uses outlined by the brief. Our designs will also draw on pattern, geometry and repetition which are central features of the Mediterranean and Islamic culture. Outdoor galleries and public spaces will be clad with colourful tiles, produced by local artisans. These poly chromatic tiles will give each space a different feel and identity and to allow for intuitional navigation within the public passages.

HOW CAN WE MAKE A MEANINGFUL INTERPRETATION OF MOROCCAN ARCHITECTURE WHICH RECONCILES TRADITION AND INNOVATION?

Functional spaces are arranged around courtyards and internal streets, forming a series of archetypes. The clustering of these archetypes creates a new promenade, a souq, providing more public space for the city in an intermediate level between the Plaza and the River.

Outdoor galleries and public spaces would be clad with colourful tiles, produced by local artisans. These poly chromatic tiles would give each space a different feel and identity.

ENVIRONMENT: WORKING WITH NATURE

Traditional Islamic architecture and urbanism is accomplished at dealing with seasonal hot and harsh climates. The building is thought of as part of an ecosystem that crucially includes other buildings. Buildings should not rely on expensive and energy inefficient technology, they should employ passive strategies that work with the environment. Each technique, SHADE, WIND and MASS, supplements each other in creating comfort not only for tenants, but also for the public realm.

The massing and urban porosity favours natural ventilation allowing prevailing wind to flow within the square, passages and small courtyards. The compact and dense fabric naturally shades the buildings and spaces between the buildings. We seek to use traditional construction and have been investigating the use of ancestral technique of mud brick walls and traditional render, implementing passive technology and increasing the thermo-mass of the buildings.

Some roofs will be accessible to public functions (hotel and workshops) when appropriate; we are also treating the flat surfaces as planted roofs to naturally increase the thermo mass of the buildings.

We propose the use of ceramic tiles to cool the public passages, using traditional technique to reduce radiant heat by evaporation and cooling effects.

HOW DO WE DEVELOP SUSTAINABLE SOLUTIONS TO DELIVER THERMAL COMFORT AND ENERGY EFFICIENCY IN THE HEAT OF THE SUN?

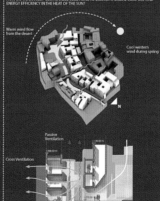

MASS
The massing, oriented from west to east, helps to generate wind flow from the plaza down to riverside. Shaded public spaces are created in between the buildings in the form of courtyards and internal streets.

WIND
Natural Ventilation through the public spaces help to maintain a comfortable climate inside the galleries, workshops and stores. The courtyards and internal streets provide shading to locals and visitors.

SUN
Tall narrow alleyways, shaded internal riads, screens, projecting bays, screened terraces and trees ensure that there is minimal solar gain throughout the year.

Whilst descending from Place Lalla Yeddouna to the riverfront, thresholds are provided where the sounds, breeze and views of the river are framed and channelled through alleys and riads. These new connections give the river back to the public. The public facilities that flow from Place Lalla Yeddouna down to the Fez River will fill the riverfront with people. Enclaves of varying sizes along the riverfront provide opportunities for different activities while also enabling a fluid pedestrian flow.

02 PLACE LALLA YEDDOUNA

010067

SITE PLAN | SCALE 1:200 @ A0

Landscape Features

Outdoor paving will be made out of limestone from Taza and a necessary gap will be introduced to create tree pits.

Outdoor benches will be made out of limestone, sourced from Taza. A gap will maintained between benches and paving.

A stone step will be used to negotiate the levels between the entrances to refurbished buildings. A gap of 100 mm will be maintained between paving and exterior wall.

The specific names of workshops will be engraved in stone above the entrances of new buildings.

TRANSVERSE SECTION A | SCALE 1:150 @ A0

04 PLACE LALLA YEDDOUNA

DISTRIBUTION OF USES

- Category A - Existing Buildings
- Category B - Existing Buildings
- Category C - Existing Buildings
- Category D - New Construction

Level -01 (River Level)

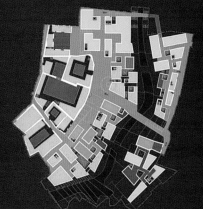

Level 00 (Entry Level)

Level 01   Level 02   Level 03

- Artisan Facilities: Workshops, Display and Direct Sales
- Public and Tourist Services: Information Centre, Retail and Exhibition Spaces
- Public and Tourist Services: Restaurants, Cafes and a Hotel
- Municipal Facilities: Management Offices, Post Office and Fire

PROGRAMME
- Artisan Facilities: Workshops, Display and Direct Sale
- Public and Tourist Services: Information Centre, Retail and Exhibition Spaces
- Public and Tourist Services: Restaurants, Cafes and Hotel
- Municipal Facilities: Management Offices, Post Office and Fire Brigade

TRAFFIC
- Public Pedestrian Routes
- Hotel Visitors and Parking Access
- Retail Spine - Souq

Direction to Zone B
Direction to Seffarine Square
Derb Jamal Chouk
Direction to Derb Ja Ou Nzel
Monumental Doors to Place Lalla Yeddouna and New Facilities
Bin Lamdoun Parking
New Riverside Promenade
Direction to Zone A

OPEN SPACES
- Primary Public Spaces: Streets, Square and River side
- Secondary Public Spaces: Urban Links and Souq
- Semi Private Spaces: Artisan's Courts, Terraces for Cafes and Restaurants

ACTIVITIES
ON PLACE LALLA YEDDOUNA

DAILY USE:
Perimeter Density
Place for meetings, retail activities and passerby

STAGE:
Sided Density
Concerts by local musicians, and other festive events

OPEN MARKET PLACE:
Central Density
Temporary open market for fruit and vegetables

10 PLACE LALLA YEDDOUNA

010067

Inner Richness
Identity, Character...

Diversity Within Unity
Modulation, Accents, Variation...

Crafts
Heritage, Tradition...

Interaction - Integration
Legibility, Navigability, Narrative of walking...

Community
Public, Social Space, Culture...

1. View from the Entrance to Medina.
2. Passage towards Place Lalla Yeddouna
3. View from Place Lalla Yeddouna, Looking South
4. Entrance to the Souq and Workshops
5. View from the Souq
6. View from the River Promenade Looking East

Medina of Fes
UNESCO World Heritage Site

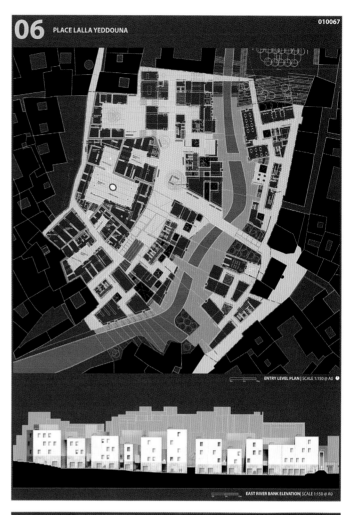
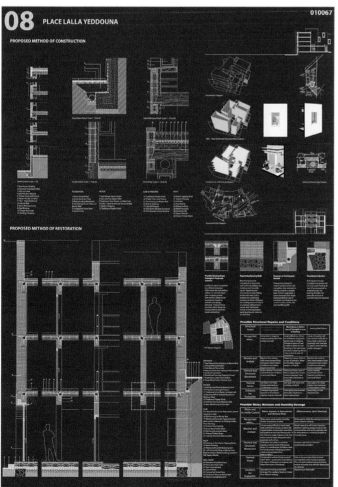
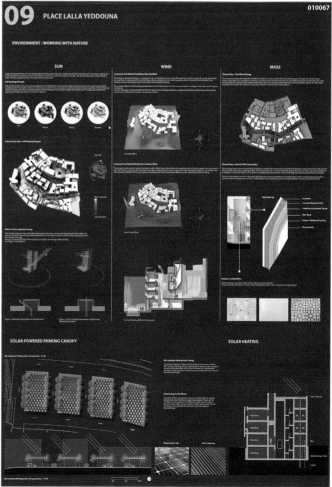

竞赛二等奖 获奖事务所

Studio Ferretti-Marcelloni

Rome (I); with Bahia Nouh, Fez (MA)

方案设计者：Laura Valeria Ferretti, Maurizio Marcelloni, Valeria Botti, Sveva Brunetti, Filippo De Dominicis, Benedetta Di Donato, Bahia Nouh
自由职业者：Matthieu Gabay, Alessandra Ienca, Valeria Leoni, Davide Palmacci, Lorenzo Senni
顾问：OSA, Munich (D); Gianluca Vanin, Mutsuko Sato

竞赛三等奖　获奖事务所

Moxon Architects

London (GB); with Aime Kakon, Casablanca (MA)

方案设计者： Ben Addy, Aime Kakon
参与者： Tim Murray, Tim Barwell, Bethany Wells, Alex Kaiser, Kyle Buchanan
顾问： Flint & Neill, London (GB), David MacKenzie; ARUP, London (GB), Gregoir Chikaher, Ian Carradice; Gardiner & Theobald, London (GB), Theo Constantinides

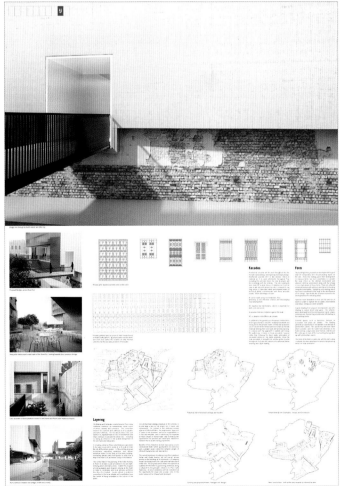
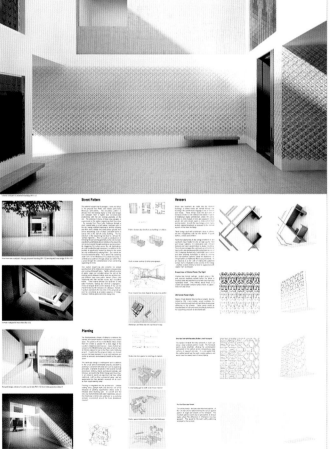

第二轮 获奖事务所

Bureau E.A.S.T.

Los Angeles (USA)/Fez (MA); with Atelier3AM, Toronto (CDN); Hashim Sarkis Studio, Beirut (RL)

方案设计者： Takako Tajima, Zineb Medaghri Alaoui, Taymoore Balbaa, Hashim Sarkis
参与者： Aziza Chaouni, Stanislav Jurkovic, Chris Wong, Hsiao Chieh, Fadi Masoud, Jeff Cogliati, Bassima Jazouili, Stephen Mauro, Karl Van Es, Melissa Cao, Youssef Halim, Carolyn Matsumoto, Metthew Brown, Amanda Chong, Nadine Koch, Penn Rudermann, Cynthia Gunadi, Lauren Martka, Maximilian Thumfart
自由职业者： Patrick Spear, Nenad Katic, Wilson Rodas
顾问： Bollinger+Grohmann, Frankfurt am Main (D) Manfred Grohmann; SETAQ Engineering, Fez (MA), Abdelai Qarqabi; Raja Mejati, Rabat (MA); Transsolar Energietechnik, Stuttgart (D), Matthias Schuler, Raphael Lafargue, Arnaud Billard, Aurelien Gervasi; Marco Schmidt Landscape Architect, Berlin (D)

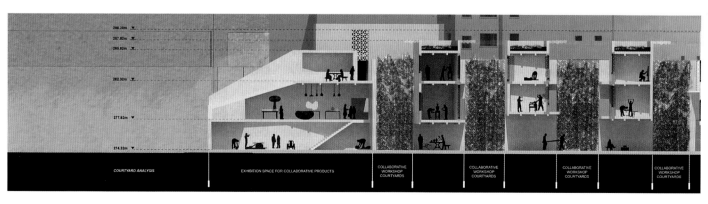

马尼托巴大学校园及周边地区

加拿大 温尼伯

2012—2013

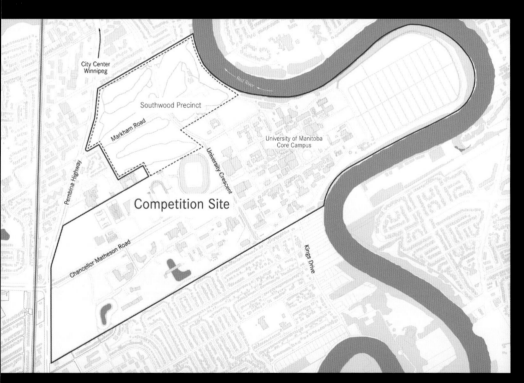

概况
业主：马尼托巴大学
项目规模：基地面积 279 公顷
类型：双阶段国际开放型设计竞赛（开放申请程序）
参赛者：第一阶段 45 组，第二阶段 6 组
竞赛预算：240 000 加拿大元（第二阶段奖金 60 000 加拿大元，第二阶段每组参赛者的费用 30 000 加拿大元）

评委
建筑评委：Marc Angélil, Architect, Zurich; Geni Bahar, Traffic Planner, North York; Ray Cole, Building Scientist, Vancouver; Jennifer Keesmaat, City Planner, Toronto; Tobias Micke, Landscape Architect, Berlin; Julie Snow, Architect, Minneapolis
专家评委：David T. Barnard, President and Vice Chancellor, University of Manitoba, Winnipeg; Kiki Delaney, University of Manitoba Alumnae, President, C.A Delaney Capital Management, Toronto; Ovide Mercredi, Misipawistik Cree Nation, Winnipeg; Scott Thomson, President and CEO, Manitoba Hydro, Winnipeg

"竞赛致力于通过一个千载难逢的机会在颇具吸引力和历史意义的区域——加里堡建造一座校园，在未来将其转换为充满生机与活力的都市村庄——这里有丰富多彩且充满活力的日常生活，有景色优美的公共空间，有极具生产潜力的地方企业，有良好、自由的学习和科研氛围。项目的目标是通过区域内的建筑设计和环境营造来提升并丰富校园体验，满足学生、教职员工以及周边社区居民的多元需求。"

——引自《竞赛摘要》

马尼托巴大学
马尼托巴大学位于加拿大马尼托巴省首府——温尼伯，是一所公立大学。该大学始建于 1877 年，有两个校区：一个是位于市中心的班纳坦校区，这里是医学院所在地；另一个是市区以外的加里堡校区，这里是宿舍和其他学生服务设施所在地。该大学共设 20 个系和 3 个学院，共有 26 000 名在校生和 6000 名教师。

加里堡校区
加里堡校区位于温尼伯市中心以南 13 千米处。校园东边以红河为界，西侧以贸易区和商务区的交通要地——彭比纳公路为界。这里与温尼伯市区类似，坐落于传统、古老的 Anishinaabe 部落的土地上并地处梅蒂斯人的居住地。280 公顷的竞赛区域包括整个现有校园及其西北部 49 公顷的 Southwood 高尔夫球场。校园附近连通新市区巴士快线和高尔夫球场用地权的获取使校园与其周边新建社区的联系更加紧密，交通也更加便捷。

竞赛任务
竞赛的关键是制订确保未来校园综合、全面发展的总体规划。上位规划针对该区域内 7 个分区的基本情况和不同特点，包括人口密度、公共基础设施、建筑形式和景观设计等，对城市用地进行混合式开发，同时确保公共交通可达性。在高尔夫球场用地区域内打造 4200 个居住单元和 21 000 平方米的零售商店及餐厅，作为校园周边新区深化设计的试点项目。竞赛中的每个流程均是有序、高效的。

马尼托巴大学校园及周边地区，温尼伯

第一阶段参赛者

第二阶段入围者

1 IAD Independent Architectural Diplomacy, Madrid (E)
2 nodo17 Architects, Madrid (E)
3 AECOM, Winnipeg (CDN)
4 Janet Rosenberg & Studio, Toronto (CDN)
5 DTAH, Toronto (CDN)
6 Perkins+Will, Vancouver (CDN)

其他参赛者

7 SBArch Architetti Associati, Rome (I) **8** OD205SL, Delft (NL) **9** Agence Up, Paris (F) **10** Mila/Jakob Tigges, Berlin (D) **11** gmp International, Hamburg (D) **12** Breimann & Bruun Landschaftsarchitekten, Hamburg (D) **13** Burgos & Garrido Arquitectos Asociados, Madrid (E) **14** C+S Architects, Treviso (I) **15** SAM Architects and Partners, Zurich (CH), with Steven Fong, Toronto (CDN) **16** Modul PKB, St. Petersburg (RUS) **17** Keith Williams Architects, London (GB) **18** Moxon Architects, London (GB) **19** Pascal's Limaçon Creative Teamwork, St. Petersburg (RUS) **20** Inbo, Amsterdam (NL) **21** Reset architecture and Buro Bol, 's-Hertogenbosch (NL)

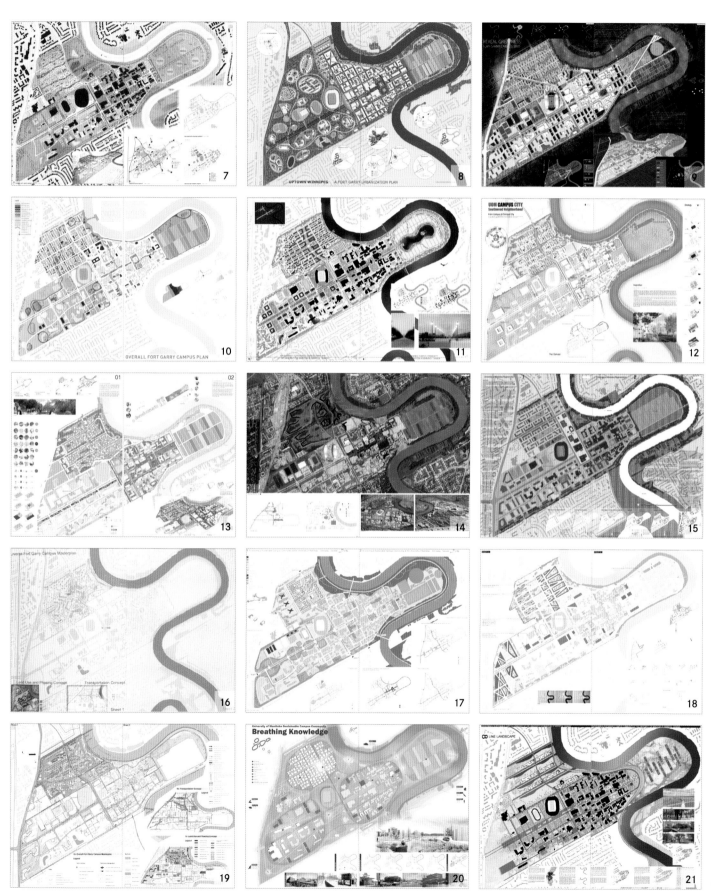

第一阶段其他参与者

22 LOHANATA Design, Jakarta (RI) **23** BVN Donovan Hill, Melbourne (AUS) **24** Didrihsons un Didrihsons, Riga (LV) **25** Creative Landscapes, Naples (I) **26** Anna Conti Architetture, Florence (I) **27** Architectural Institute of British Columbia, Vancouver (CDN) **28** TD, Flachau (A) **29** Campus 2 Cover, Moralzarzal (E) **30** TRYS A.M. Architects, Vilnius (LT) **31** caramel architekten, Vienna (A) **32** Shahla Shahmoradi, Tehran (IR) **33** The Commons/L'OEUF, Montreal (CDN) **34** ECOTONelu Design Studio, Seongnam (ROK) **35** MMM Group Limited/Number TEN Architectural Group, Ottawa (CDN) **36** Plain Projects, Winnipeg (CDN) **37** FT3 Architecture Landscape Interior Design, Winnipeg (CDN) **38** CIVITAS Urban Design and Planning, Vancouver (CDN) **39** B+H Architects, Toronto (CDN) **40** ContisNambiar, Brooklyn (USA) **41** DUDA/PAINE Architects, Durham (USA) **42** Brown and Storey Architects, Toronto (CDN) **43** 5468796 Architecture, Winnipeg (CDN), with Atelier Anonymous, Vancouver (CDN), Aziza Chaouni Projects, Toronto (CDN), and Hilderman Thomas Frank Cram Landscape Architecture & Planning, Winnipeg (CDN) **44** NOA Architecture, New York (USA) **45** RSE Landscape Architecture, Amsterdam (NL)

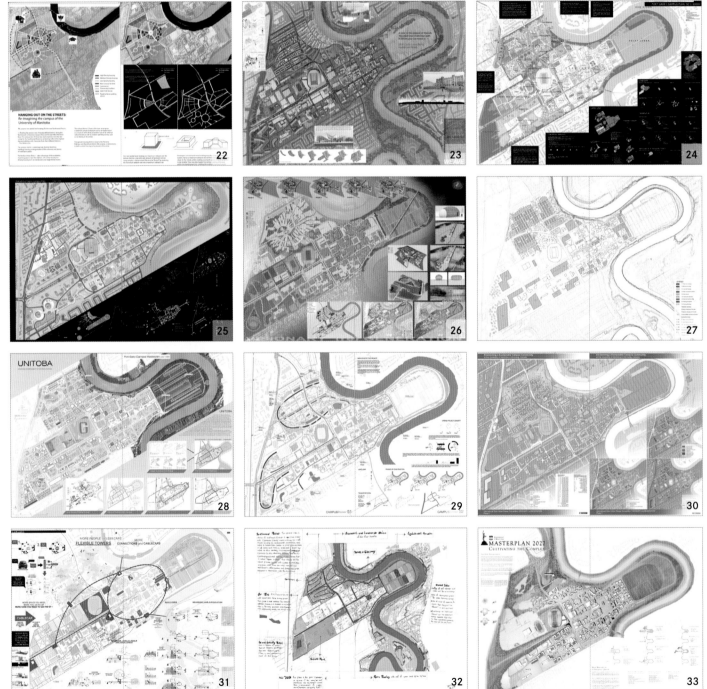

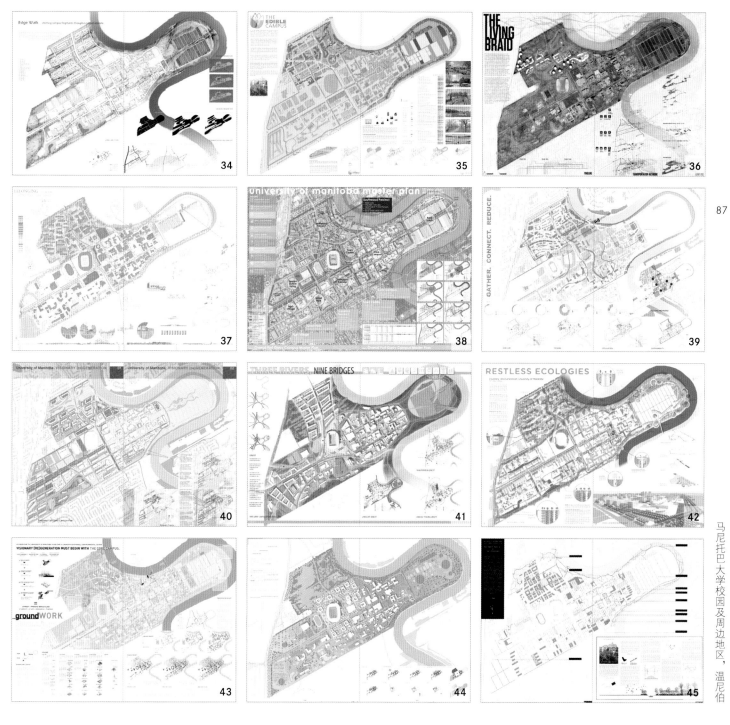

第二阶段参赛者

获奖者

1 一等奖   Janet Rosenberg & Studio, Toronto (CDN)
2 二等奖   Perkins+Will, Vancouver (CDN)
3 三等奖   DTAH, Toronto (CDN)
4 四等奖   IAD Independent Architectural Diplomacy, Madrid (E)
5 第二轮   nodo17 Architects, Madrid (E)
6 第一轮   AECOM Canada, Burnaby (CDN)

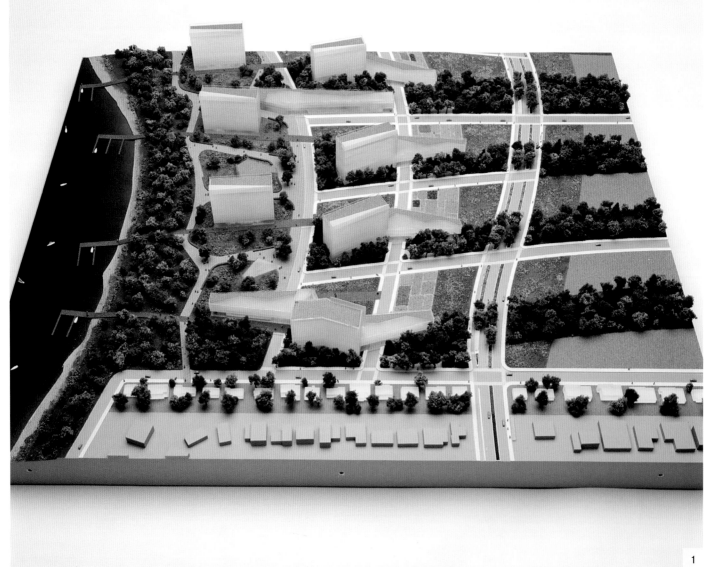

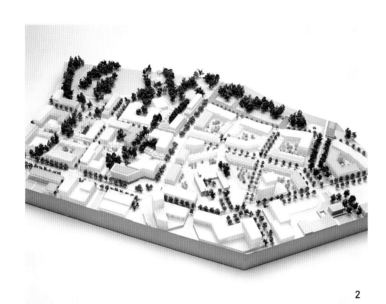

2

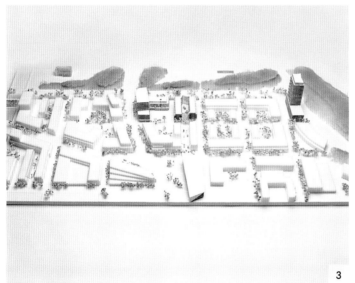

3

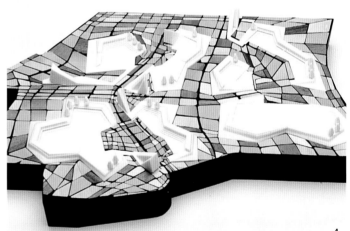

4

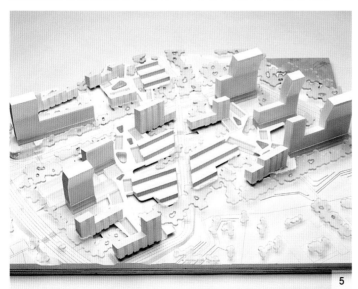

5

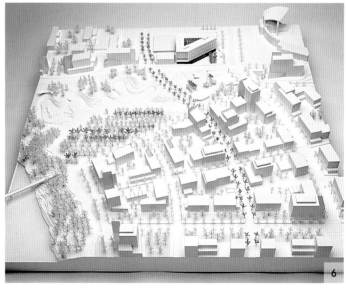

6

竞赛一等奖　获奖事务所

Janet Rosenberg & Studio

Toronto (CDN); with Cibinel Architecture, Winnipeg (CDN);
Landmark Planning & Design, Winnipeg (CDN)

方案设计者：Janet Rosenberg, George Cibinel, Ateah Curwood, Donovan Toews
参与者：Glenn Herman, Maury Mitchell, Todd Douglas, Jenny Bukovec, Justin Miron, Reinaldo Jordan, Jessie Seed, Joseph Orobia, Matt Cibinel

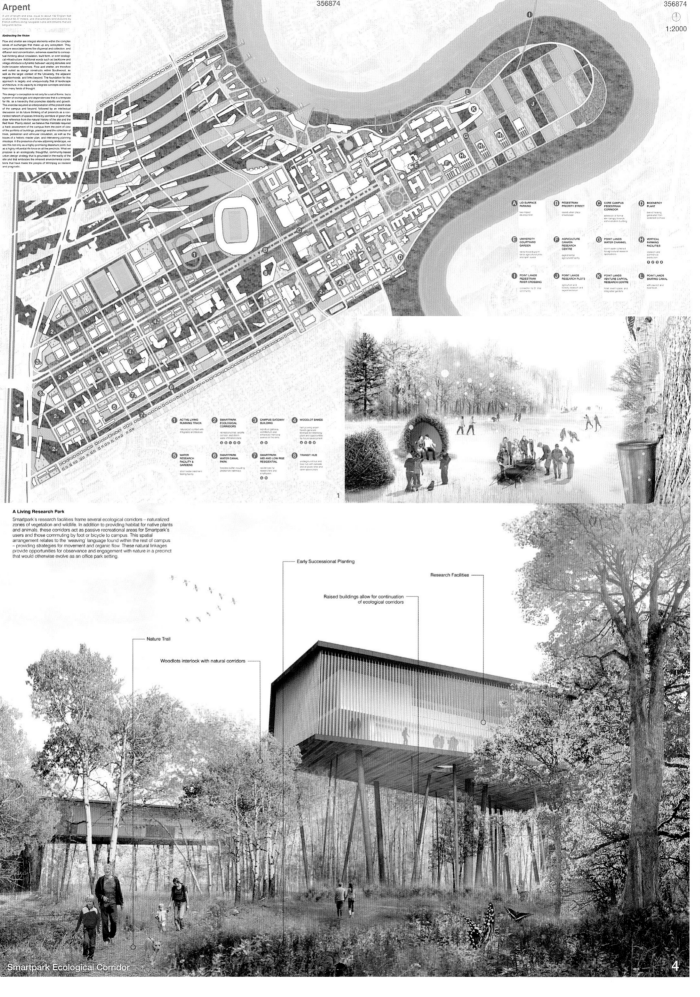

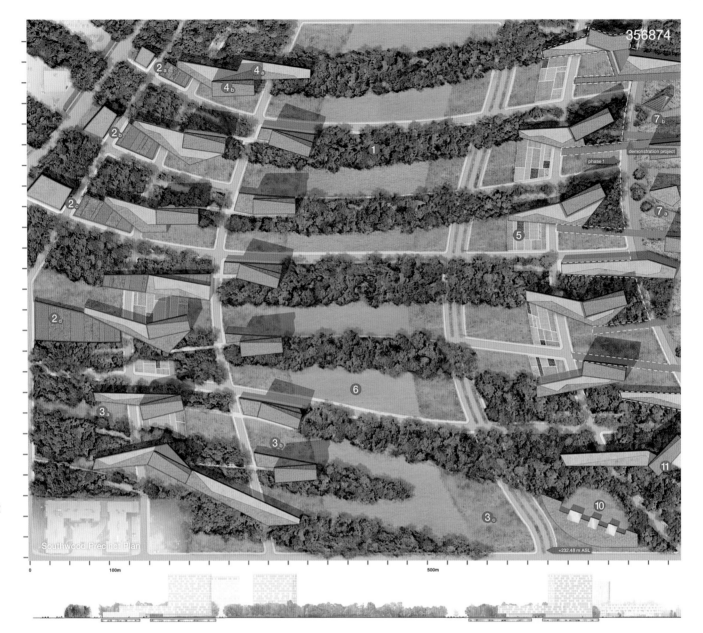

Southwood Precinct Plan

1. **SOUTHWOOD FOREST BANDS**
   extend beyond site boundary creating linkages with context
2. **SOUTHWOOD PEMBINA COMMERCE** a b c d
   large commercial locations with southwood and local clientele
3. **SOUTHWOOD TALL GRASS PRAIRIE OPEN SPACE ROAD BUFFER** a b c
   naturalized open space processional axis
4. **SOUTHWOOD BUILDING TYPOLOGY** a b
   4-storey podium & 12-storey tower residential & commercial blocks
5. **SOUTHWOOD COMMUNITY GARDENS**
   dispersed plots for residents and local community
6. **SOUTHWOOD FAIRWAY LAWN**
   passive play non-programmed active living open space
7. **SOUTHWOOD RIVERFRONT BOARDWALK PAVILIONS** a b
   open space viewing platforms with research facilities and rentals
8. **SOUTHWOOD RIVERFRONT BOARDWALK**
   open space with islands of riparian ecology
9. **SOUTHWOOD TERMINAL**
   local bus transit hub and parking structure
10. **SOUTHWOOD ENVIRONMENTAL MAGNET SCHOOL**
    supporting southwood and campus community
11. **CONTINUING EDUCATION BUILDING**
    including community health access centre, plaza, and retail
12. **RIVERBOTTOM FOREST**
    enhanced riparian vegetation zone
13. **SOUTHWOOD BOATHOUSE**
    boat storage, rentals, community programs, and river view roof terrace facilities

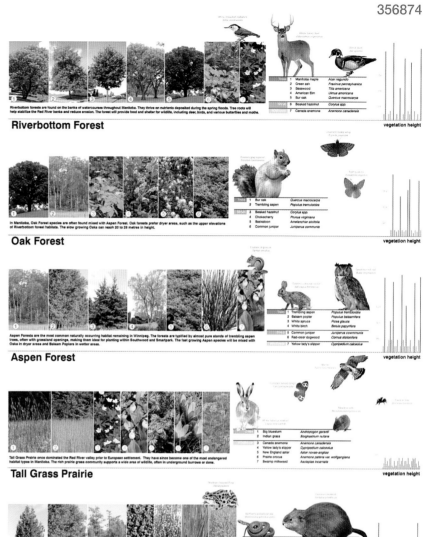

Riverbottom Forest

Oak Forest

Aspen Forest

Tall Grass Prairie

Wetland

Southwood Riverfront Boardwalk

Providing abundant space for active living and informal gatherings, the riverfront boardwalk is a meeting place with riparian islands, providing habitat for native plant and animal species. Built elements of natural materials sensitive to the river ecology ground the observer in the spirit of the place. Responding to the natural curvature of the river, curvilinear benches provide opportunities for observation and reflection.

竞赛二等奖　获奖事务所

Perkins+Will

Vancouver (CDN); with 1X1 Architecture, Winnipeg (CDN),
PFS Studio, Vancouver (CDN)

方案设计者：Ryan Bragg, Joyce Drohan, Glen Gross, Kelty McKinnon, Chris Phillips
参与者：Achim Charisius, Jamuna Golden, Catarina Gomes, Krisan Osterby, Adam Slawinski, Ben Sporer, Yong Sun, Travis Cooke, Jason Kun, Markian Yereniuk, Lin Lin, Jenna Buchko
自由职业者：Trevor Butler, Chris Foyd, Gordon Harris, Richard Littlemore

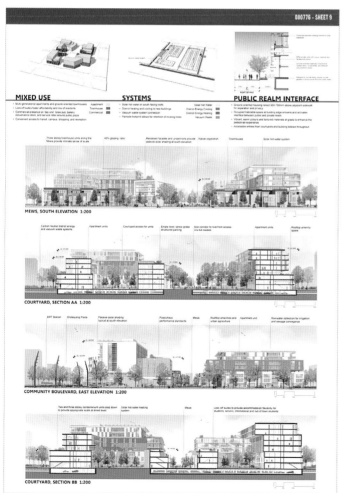
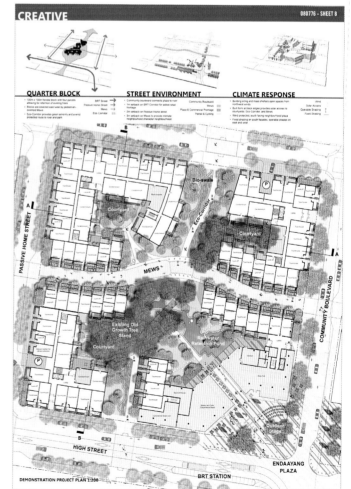

竞赛三等奖　获奖事务所

DTAH

Toronto (CDN); with Cohlmeyer Architecture Limited, Winnipeg (CDN);
Integral Group, Winnipeg (CDN)

方案设计者：Robert Allsopp, Steve Cohlmeyer, Integral Group

SOUTHWOOD PRECINCT PHASE 1

Phase 1 establishes the key organizing elements of the new Southwood residential neighbourhood, including the new Esplanade, Central Square - Gathering Place, and the Integrated Storm Water and Open Space System. The first phase brings a rich mix of residential building and unit types, retail and community uses, semi-public and public spaces. The northern blocks of the new village establish a strong character along the Commons edge and introduce the first phases of infrastructure associated with the district energy system for the Village 'Ambient Loop'.

Southwood Precinct Phase 1 Site Plan 1:1000

1. Southwood Public Building - The Forum
2. Apartment Blocks with Atrium Courtyard
3. Apartments with Grade-Related Units
4. Mixed Use Residential with Commercial-at-Grade
5. Townhomes with Rear Parking
6. Retail, Commercial and Hospitality
7. Living Machine Interpretive Centre
8. Central Square - The Gathering Place
9. Southwood Esplanade
10. Residential Courtyards
11. The Canal
12. Storm Water Pond
13. The Commons
14. Woonerf - Shared Street
15. Stadium Forecourt
16. Retained Tree Stand
17. Earth House
18. Floating Canoe Dock - Gateway to the Red

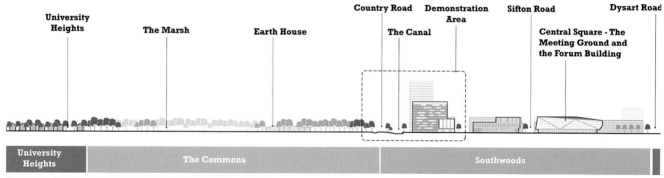

University Crescent Elevation - New Campus Frontage 1:1000

View of Residential Green Street

Southeast View of The Meeting Ground

竞赛四等奖　获奖事务所

IAD | Independent Architectural Diplomacy

Madrid (E)

方案设计者： Stéphane Cottrell, Jér.me Michelangeli, Rafael Sà
参与者： Ignacio Tellado, Daniel Olmo, Sara Garcia, Jurgita Mockute, Jack Marston

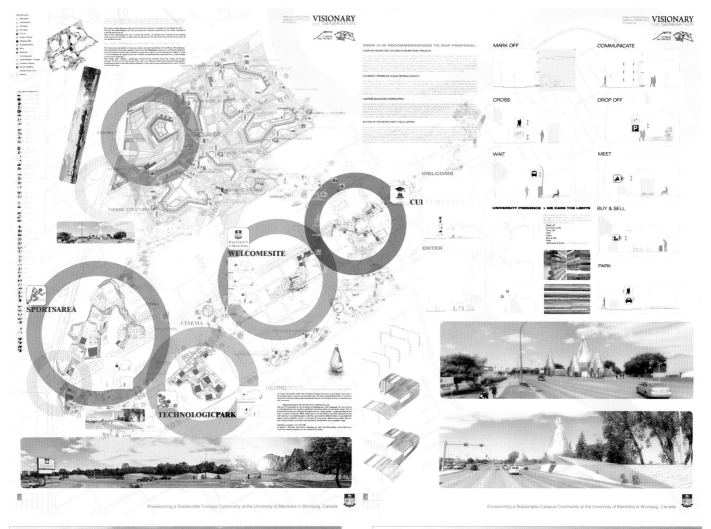

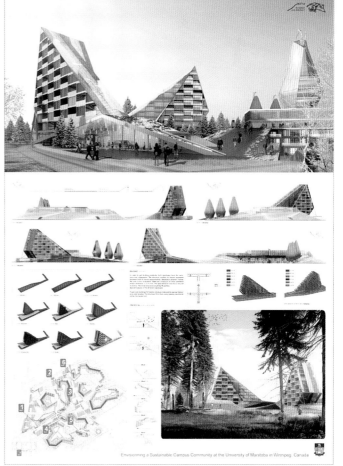

越南 – 德国大学校园

越南 胡志明

2012—2013

概况
业主：越南社会主义共和国（以越南教育培养部为代表，与世界银行合作）
项目规模：基地面积 50.5 公顷
类型：开放型设计竞赛（根据世界银行的相关指导原则，设计招标由总承包商组织，各参赛者公开投标）（开放申请程序）
参赛者：5 组
竞赛预算：150 000 美元

评委
建筑评委：Prof. Donald Bates, Architect, Melbourne; Prof. Gulzar Haider, Architect, Lahore; Prof. Zvonko Turkali, Architect, Frankfurt am Main; Dr. Tran Thanh Binh, Architect, President of Institute for Research and Design of Schools, Hanoi; Khuong Van Muoi, Architect, Vice Chairman Architecture Association, Ho Chi Minh City
专家评委：Prof. Bui Van Ga, Ph.D, Deputy Minister, Ministry of Education and Training of Vietnam (MOET); Eva Kühne-Hörmann, Minister of Higher Education, Research and the Arts, State of Hesse, Germany; Prof. Dr. Jürgen Mallon, President of Vietnamese-German University (VGU); Nguyen Dinh Toan, Deputy Minister of Ministry of Construction (MOC) of Vietnam

"教育、科研和技术设施应该彰显德国品质并达到德国标准。校园设计应该在满足上述要求的基础上反映越南的气候条件、历史文化、居民生活习惯。

"可持续发展"是校园设计的总体原则——对区域自然资源及历史人文资源等进行全方位整合(包含但不限于生态资源),力求做到:方案设计合理、全面,具体实施过程可操作性强且有序、高效。"

——引自《竞赛摘要》

越南-德国大学

越南政府对高等教育系统的一系列改革是贯彻"可持续发展"的一项重要举措。该项目提出了新的大学建设计划——大学应该让年轻一代在专业对口且符合国际标准的环境中得到锻炼。改革的第一步是建立四所"示范大学",每所大学与一个国家合办。其中之一是越南-德国大学。合作伙伴国在课程建设方面提供支持,并就结构设施设计和实施的相关事宜向教职员工及工程人员提供指导和咨询服务。该项目由德国联邦教育与研究部、海塞省以及部分高等学府提供技术支持;项目经费多半来自世界银行的贷款。

场地

竞赛的选址位于越南平阳省北部胡志明市西北方向约 50 千米处。这里是越南全国发展速度最快的地区之一,拥有工业综合体、科研机构和大型住宅。将近 51 公顷的竞赛场地位于越南国家 13 号公路(HCMC 主干道),北至四环路。现有的乡村住宅形成了基地南侧边界。场地中设有排水渠。

竞赛任务

大学将分两期建设。第一阶段(2017 年前)总基地面积约为 13.5 万平方米,供 5000 名学生使用,第二阶段增加 19 万平方米的基地面积,专供 12 000 名学生使用。竞赛任务是通过举办竞赛来确定新校园的总体规划。除了用于教学、科研和技术创新的公共设施外,在校园内还将建造运动场并提供与之配套的各种运动设施,同时为学生及教职员工提供住房及住宿设施。方案设计需考虑当地的地理、气候和文化因素。

越南-德国大学校园,胡志明

参赛者

1 一等奖 Machado and Silvetti Associates, Boston (USA)
2 二等奖 KSP Jürgen Engel Architekten, Frankfurt am Main (D)
3 三等奖 Itsuko Hasegawa Atelier, Tokyo (J)
4 四等奖 Riegler Riewe Architekten, Graz (A)
5 五等奖 Henn Architekten, Munich (D)

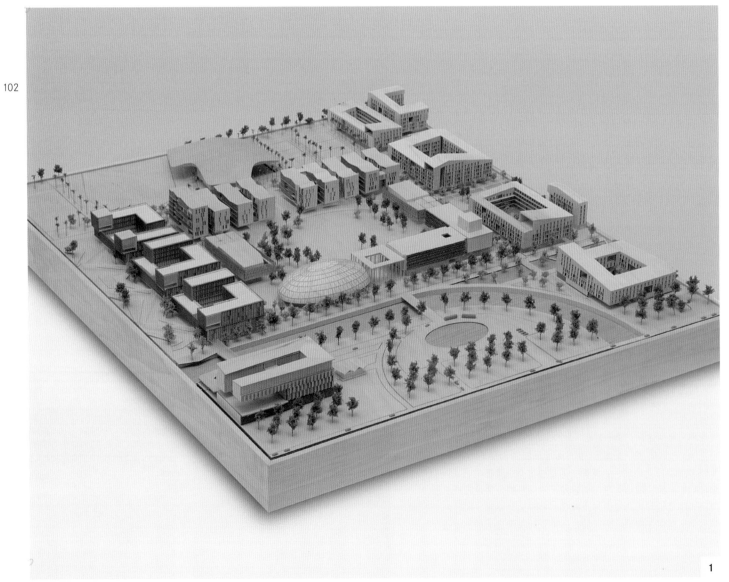

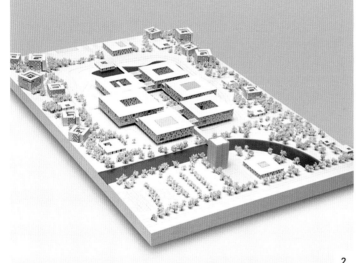

2

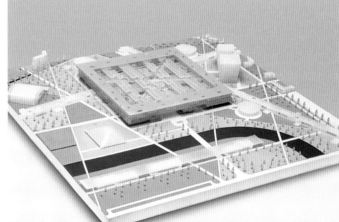

3

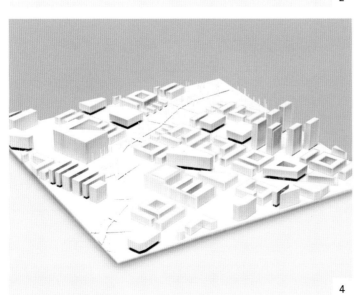

4

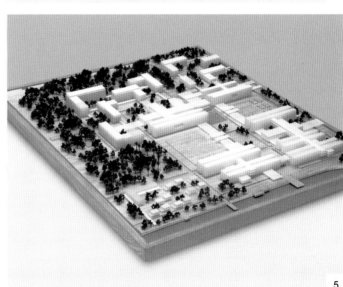

5

竞赛一等奖 获奖事务所

Machado and Silvetti Associates

Boston (USA); Happold Ingenieurbüro, Berlin (D)/London (GB);
Levin Monsigny Landschaftsarchitekten, Berlin (D)

方案设计者：Jeffry Burchard, Evan Brinkman, Carly Lamb
参与者：Victoria Gigena, Lauren Di Pietro, Jose Ribera, Gabriel Tyszberowicz, Christian Lavista, Janet Bacovich, Dany Gutierrez, Nico Viterbo
自由职业者：Dongwoo Yim, Rafael Luna, Carmine D'Allessandro, James Carrico, Yu Chen, Paul Fiegenschue
顾问：Arup, Cambridge (GB)/Ho Chi Minh City (VN); Grant Associates, Singapore (SGP); Hoang Giang Construction Consultant, Ho Chi Minh City (VN)

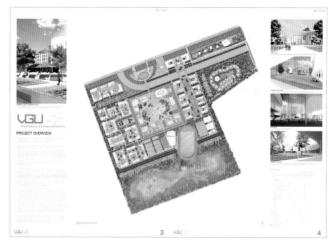
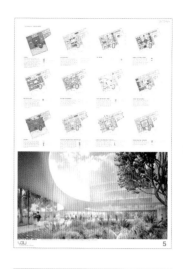

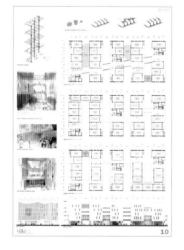

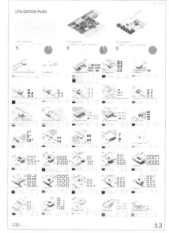
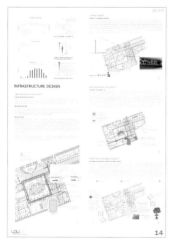

RESIDENTIAL COURT & COMMUNITY CENTER
A small club-like Community Center, or Everyone's House. It is a direct homage to the vernacular domestic imagery of Vietnam, a one story building of rooms, courtyards and galleries so precisely designed that the building becomes a culture-specific piece of architecture.

Vietnamese-German University

PROJECT OVERVIEW

We have opted to design a dense and compact campus, one that does not utilize or spill over the entire available land area but leaves a portion of the site open and available for enjoyment as a natural space, or can be developed in the future as a campus annex or for campus expansion.

This strategy has three major advantages: First, it produces great economy in campus construction and implementation, generating savings in road, infrastructure and maintenance costs. Second, it facilitates pedestrian access to and from all buildings in an efficient way, shortening time spent in human movement while increasing casual interaction between students and faculty. This design makes for a more collegial daily experience as well as a richer campus life and fosters the practice of a civilized academic urbanity. Third, it creates a campus with strong edges and clear boundaries between itself and the surrounding neighborhoods, a campus that is thus easy to identify and secure. Above all, this compact strategy results in a campus with a strong formal presence, a strong image and recognizable identity, a campus that functions like an academic town clearly occupying and defining an urban district in a city, or having a commanding presence in a suburban or rural landscape. We like to think of this campus as an "object", an objectified piece of campus fabric.

The design of this campus is based on a simple urban design /architecture concept: The ground level will be a continuous garden as free, accessible and traversable in all directions and as transparent as possible. We think of it as a porous 4m tall layer of campus social space occupied only by enclosed spaces and rooms (islands in a sea of green) dedicated exclusively to communal activities and communal functions; this space is punctuated by the columns of the buildings above, and its generously furnished with built in tables and bench as well as garden furniture. The ground plane can also be thought of as a teaching botanical garden from which faculty, students and staff can derive pleasure and relaxation. A continuous, 4m tall canopy encircles and forms The Commons at the center of campus. This canopy is pulled and stretched in all directions, thus providing continuous cover and protection from rain and sun to all pedestrians moving from one end of the campus to the other, a perfect solution for the local climate.

As a unique institution consolidated from the intersection of the two diverse cultural traditions, the Socialist Republic of Vietnam and the Federal Republic of Germany, the VGU seeks to accomplish the difficult task of incorporating the unique qualities of both, be they geopolitical, linguistic, social, climatic or any of the other myriad traditions rooted in each culture. We were inspired by the Vietnamese idea of a continuous garden, a lush paradise of exuberant vegetation and a plethora of flowers. We used Vietnam's domestic color palette and its rich vernacular building traditions and took inspiration from the grand planting of its French Colonial parks, the assertiveness of its public edifices and more importantly, the remarkable stock of modern buildings, from approximately 1930 to 1950 found in Ho Chi Minh City. We were similarly inspired by the German idea of efficiency, systematic rigor, methodical organization with rational order, contemporaneity, legibility and intellectual depth ever present in German thought and products. More concretely, the many recent educational buildings seen in respected universities in Hessen, for instance, motivated us because the most advanced pedagogies are being implemented inside those technologically advanced, yet sometimes generic buildings.

It does seem ideal to us to work with both traditions, with the controlled overlapping of the sensual and the intellectual dimensions they contain. This will not only lead to the construction of a unique campus now but, in cultural terms, this rich mode of working will produce environments which will contribute to the formation of future cosmopolitan leaders in science and technology.

GROUND PLAN WITH ROOF OUTLINE SHOWN
Scale 1:1000

ENTRY COURT
A symbolic entrance space, a "postcard ready" campus gate. A view of the University Commons, Lecture Hall and Sports facilities beyond.

ACADEMIC QUADRANGLE
Lined by classic arcades, this open space promotes the sharing of intellectual knowledge among the various academic clusters.

FOOD COURT
A pleasant shaded public place for eating throughout the day.

UNIVERSITY COMMONS
Located at the heart of the VGU campus, this public amenity is surrounded by communal programs including the Lecture Hall, Food Courts, and Library. The effect is akin to a Central Park, where students, faculty and visitors mingle and play.

BUILDINGS

#	Building Name	Form A#	Floors	Building Foot Print (on ASL)	Building Height (m ASL)	Building Height (m Above Foot Print)
1	Entry Court	na	1	6.0	24.0	18.0
2	Exhibition Hall	1.1	1	5.5	20.5	15.0
3	Research Pavilions	1.2	1	5.5	14.5	9.0
4	Academic Club	1.3	1	5.5	11.5	6.0
5	Administration Building	2.1 (+2.2, 2.3, 2.5, 2.6, 3.1, B.1)	5/7	5.5	34.5	29.0
6	Ceremony Hall	2.7	1	6.0	26.0	20.0
7	Foundation Year	2.4 (+3.1)	6	8.5	32.5	26.0
8	Cluster I - Electrical Engineering and Information Technology	2.8 (+3.1)	5	7.0	29.0	22.0
9	Cluster II - Civil Engineering	2.9	5	7.0	29.0	22.0
10	Cluster III - Biotechnology	2.10 (+3.1)	6	6.5	32.5	26.0
11	Cluster IV - Economics and Industrial Engineering	2.11	5	7.0	29.0	22.0
12	Cluster V - Mechanical & Process Engineering	2.12	6	6.0	32.0	26.0
13	Cluster VI - Computer Science	2.13 (+3.1)	5	7.0	29.0	22.0
14	Cluster VII - Natural Sciences	2.14 (+3.1)	7	6.0	36.0	30.0
15	Lecture Hall / Study Center	2.15 (+3.1)	5/7	7.0	33.0	26.0
16	Library / Media Center	3.16 (+3.1)	4/7	6.0	32.0	26.0
17	Food Court / Cafeteria	3.1	4	6.5	23.5	17.0
18	Sports Hall	3.3 (+3.1)	2	7.5	31.5	24.0
19	Dormitory 1	4.1 (+3.2)	6/7	6.0	33.0	27.0
20	Dormitory 2	4.1 (+3.2, 3.2)	6/7	7.0	34.0	27.0
21	Dormitory 3	4.1 (+3.1)	4.5	7.7	29.7	22.0
22	Dormitory 4	4.1 (+3.1)	8	7.7	39.7	32.0
23	Dormitory 5	4.1 (+3.1)	8	7.7	43.7	36.0
24	Dormitory 6	4.1 (+3.1, 3.2)	8	7.0	43.0	36.0
25	Dormitory 7	4.1 (+3.2)	8	6.5	42.5	36.0
26	Dormitory 8	4.1 (+3.1, 3.2)	6/7	8.0	33.0	27.0
27	Dormitory 9	4.1 (+3.1, 3.2)	8	6.0	36.0	30.0
28	Dormitory 10	4.1 (+3.1)	8	5.5	41.5	36.0
29	Dormitory 11	4.1 (+3.1)	6/7	5.5	35.5	30.0
30	Dormitory 12	4.1 (+4.3)	4.5	5.5	25	20.5
31	Community Center	4.2	1	6.5	12.5	6.0
32	Academic Staff Housing	5.1	2	8.0	15.0	8.0
33	Guest Housing	5.2	3	6.5	18.5	12.0
34	Gym & Community Room	3.3	1	6.0	13.0	7.0
35	Infrastructure Building	4.1, A.2	2	7.5	27.5	20.0
36	Water Tower	A.3		7.5	27.5	20.0
37	Parking Garage	B.1	4	7.0	22.0	15.0
38	Motorcycle Parking Garage	B.1	3	6.0	16.0	10.0

▲ Main Building Entrance
▲ Service Building Entrance

LIBRARY AND MEDIA CENTER
Glowing from across the University Commons. The Vietnamese-German University at night.

LIBRARY / MEDIA CENTER

The library is part of a trio of important public buildings including the Ceremony Hall and the Entry court. These act like one large front facing the NL-4 Road towards the North and the University Commons towards the South, while producing a memorable and defining images of the VGU campus.

The form of the library is orthogonal and well-behaved and like many buildings on campus gains its identity through its particular facade, in this case an exquisite brisole screen.

LIBRARY - ENTRY COURT - CEREMONY HALL

ENTRY SEQUENCE DIAGRAM

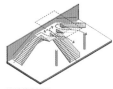
GRAND STAIR DETAIL

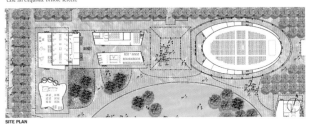
SITE PLAN

HORIZONTAL CIRCULATION DIAGRAM

SEMI EXTERIOR SPACES DIAGRAM

VOIDS DIAGRAM

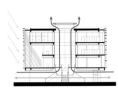
VENTILATION THROUGH BUILDING

PHASE 2
Phase 2 follows the same logic of multiple bars in order to create visibility across the campus. At the same time, it follows a logic for the master plan of raising opposite corners of the courtyard in order to anchor the space between the library tower and Phase 2.

THE LIBRARY SCREEN VISUAL EFFECTS

DETAIL VIEW OF THE FACADE FROM THE UNIVERSITY COMMONS CANOPY
A modular concrete brisole provides a contemporary building image and protects books and users from the intense sun.

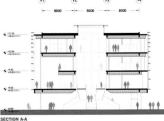
SECTION A-A
Scale 1:200

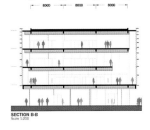
SECTION B-B
Scale 1:200

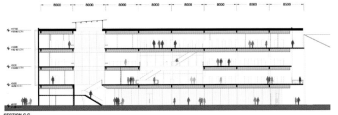
SECTION C-C
Scale 1:200

BUILDING DESIGN

We propose a set of campus buildings that are rationally different from each other. For instance, in reinforcement of the campus zoning strategy, all residential buildings are similar (with size differences, more notably), as are all academic buildings. Special, unique buildings such as the library, classrooms and ceremonial buildings, also differ from one another.

The reason for this difference is simple: we believe that campus buildings should look, or allude to what they are, that they should be legible and that they should represent their function. Over time, images and functions have merged in such a way that one expects administrative buildings to be formal or imposing, athletic buildings to be playful, structurally muscular and to display advanced roof technologies, food courts to be open and inviting, residence halls to have domestic type fenestration, etc. All that referential impetus is to be accomplished within limits so as to keep the VGU campus from becoming chaotic or confusing. In this case, the limits will be provided by our design referring only to modern and contemporary architecture, by the selection of structural systems, material finishes and window systems that will be common to most buildings on campus.

There is also a certain efficiency and economy of means present in all building designs: once the building type and image have been determined for a given function, we looked for the most economic way to build that building. The captions on this board, accompanying the drawings, describe these choices in some detail.

FROM THE ADMINISTRATION BUILDING TOWARD THE VIETNAMESE-GERMAN UNIVERSITY CAMPUS
A Presidential View over the campus core. From Left to Right: Academic Quadrangle, with the Field of Dreams in the distance. Academic Clusters. Lecture Hall / Study Center. Ceremony Hall. Entry Court. Library, with Dormitories in the Background.

MODULAR STRUCTURAL SYSTEMS
Many buildings use simple concrete columns and slabs, on regular structural bays. Buildings that are larger can easily grow bays horizontally or vertically by adding slabs. Simple spans reduce cost and improve construction times.

MODULAR BUILDING FORMS
The Modular Structural System is readily and efficiently applied to bar, courtyard and L-shape forms on the VGU campus.

FACADES
These well-known formal types are treated with programmatically appropriate facades. The facades are designed to shade the buildings throughout the year and reduce mechanical cooling loads.

SPECIAL BUILDING FORMS
A few buildings fall outside of this widely applied paradigm. The Ceremony Hall and Sports Center use long span concrete and steel construction to provide clear floor space for large events.

ACADEMIC CLUSTERS

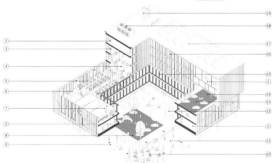

ACADEMIC COURTYARD AND ARCADE
Vibrant landscapes and furniture actively encourage lively intellectual exchanges and collaborative events.

TYPICAL ACADEMIC CLUSTER BUILDING

STUDENT DORMITORIES

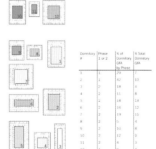

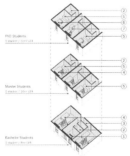

RESIDENTIAL SQUARE FROM GARDEN TERRACE
Sprinkled throughout the Residential zone, terraces like this one provide places for small group events and gatherings, people watching or catching up on the world news.

STUDENT DORMITORY MAP
Various building sizes populate the Residential zone. This variety, all produced by modular and repetitive construction and heightened through the use of color, provides a diverse set of living choices for students at VGU.

MODULAR DORMITORY LIVING UNITS
The dormitories are organized on a modular bay. This allows for efficient construction and provides extensive flexibility as the university adjusts their housing needs if the composition of the student body ever changes.

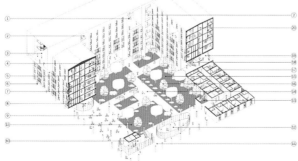

TYPICAL ACADEMIC CLUSTER BUILDING

SITTING NOOK IN RESIDENTIAL ZONE
An intimate place to meet friends, read a book, or study for an exam.

竞赛二等奖 获奖事务所

KSP Jürgen Engel Architekten

Frankfurt am Main (D)

方案设计者： Jürgen Engel, Luc Monsigny
参与者： Johannes Reinsch, Christopher Hammerschmidt, Nguyen Huy, Le Man Tien, Christoph Cellarius, Jan Klein, Zhao Xing, Jorge Veira Rob Grotewal, Tim Stawitzke, Pauline Barral

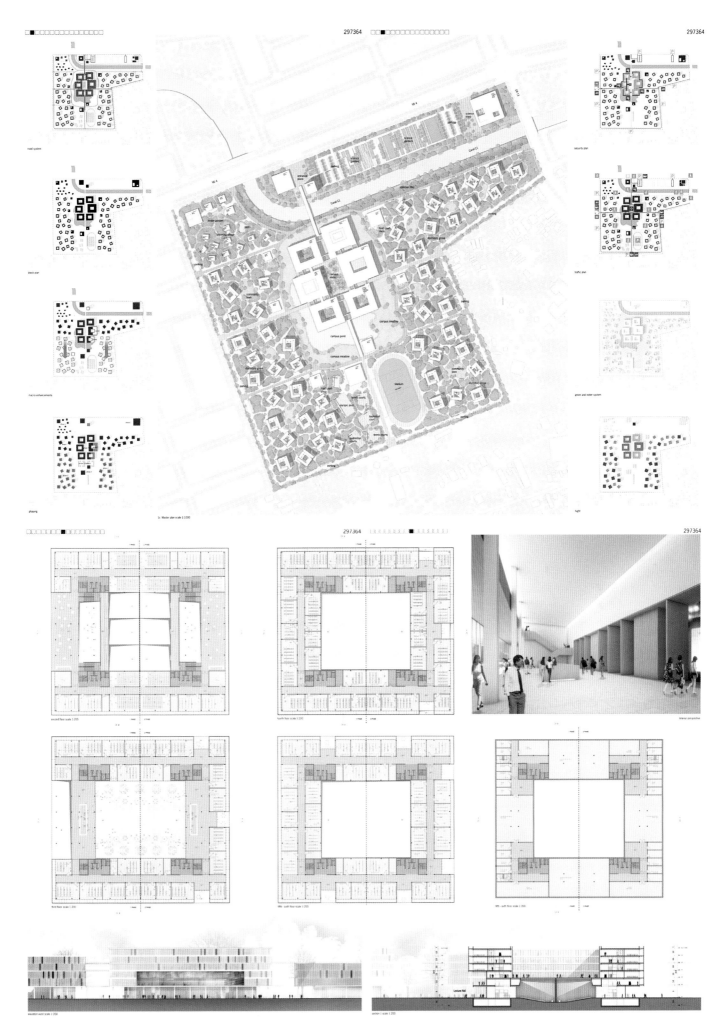

竞赛三等奖　获奖事务所

Itsuko Hasegawa Atelier

Tokyo (J); with Ove Arup and Partners, Tokyo (J); P.T. Morimura & Associates, Tokyo (J); Sirius Lighting Office, Tokyo (J); Taylor Cullity Lethlean, Victoria (AUS); Tokyo Institute of Technology, Hamamatsu (J)

方案设计者：Itsuko Hasegawa

library interior and book tower

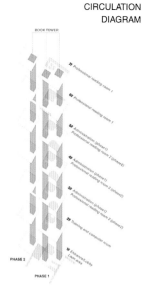
CIRCULATION DIAGRAM

1F   2F   3F/4F/5F

library plans 1/200

Book Tower

Traditional Bookshelf and core area will account for the majority of the floor. The reading spaces can often be the residual leftover space and therefore not very efficient or practical. The symbolism of the library is visualized in the elevation by the integrated books. In addition, it also serves as a display for recommendations and new books signage.

1F
The first floor serves as main Lobby, offering a generous entrance. Loan area is located in this floor given its easier access.

2F
Training room and Computer pool location. These spaces are enclosed in a free-form surface tent membrane. The sound insulation and the lighting control systems create an environment where one can focus on meeting and computer work.

3F / 4F / 5F
Administration will be located here for Phase 1 along with a series of flexible meeting spaces. These three floors will change to Professional Reading Room in Phase 2.

In combination with two upper floors, the Professional reading spaces will eventually consist of five floors.

6F / 7F
Professional reading location. A variety of furniture configurations are used in the Reading space. Please note that the Single Work Rooms and Group Work Rooms are located along the outer walls to create a quiet learning environment as a whole.

library interior and book tower

6F   7F

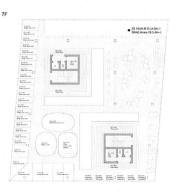
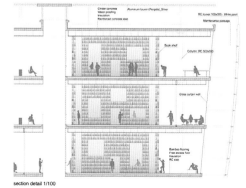

section detail 1/100

竞赛四等奖　获奖事务所

Riegler Riewe Architekten

Graz (A); with ASTOC, Cologne (D); Marina Stankovic Architects, Berlin (D); YO2 Architects, Seoul (ROK); Ingenieurbüro Dr. Binnewies Ingenieurgesellschaft, Hamburg (D); Amstein+Walthert, Karlsruhe (D)/Zurich (CH); Agence Ter.de Landschaftsarchitekten, Karlsruhe (D); Transsolar Energietechnik, Munich, (D)

方案设计者：Roger Riewe, Forian Riegler, Ingo Kanehl, Marina Stankovic, Tobias Jortzick, Young Joon Kim, Henri Bava
参与者：Anna Kollegger, Anna Nowy, Neira Mehmedagic, Alexia Eberl, Fatlum Rodoniqi, Marietta Pagonis, Manuel Hauer, Oliver Ernst, Sebastian Wahlstr.m Klampfl, Katja Loncar, Ik Ran Shin, Ju Hyun Lee, Sang Hoon Back, Dong Woo Kang, Joo Hee Han, Sung Hyun Ahn, Seung Ae Yoo, Bj.rn Dittrich, Joachim Ehrmann, Kirsten Schomakers
自由职业者：Beatrix Redlich, Susanne Gerstberger

竞赛五等奖　获奖事务所

Henn Architekten

Munich (D); with Arup, London (GB); Topotek 1 Landschaftsarchitekten, Berlin (D)

方案设计者：Christian Bechtle, Martin Henn, Martin Rein-Cano
参与者：Güley Alag.z, Paul Langley, Haitao Long, Jeewon Paek, Sascha Posanski, Klaus Ransmayr, Matthias Kolle, Janka Paulovics, Silvia Bachetti
自由职业者：Ba Hoang Tu Ly

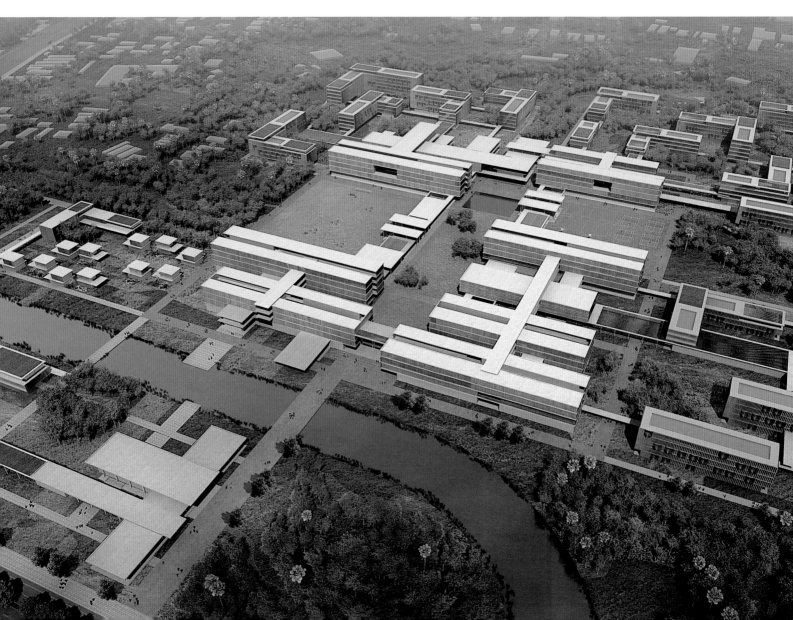

RESIDENTIAL UNIT LAYOUT

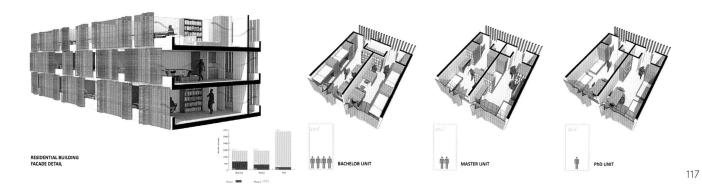

RESIDENTIAL BUILDING
FACADE DETAIL

BACHELOR UNIT   MASTER UNIT   PhD UNIT

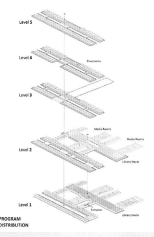

PROGRAM
DISTRIBUTION

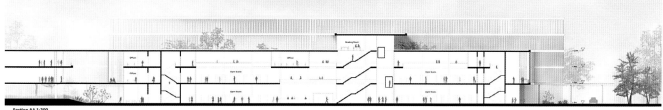

Section AA 1:200

Elevation South 1:200

河内科技大学校园

越南 河内

2013—2014

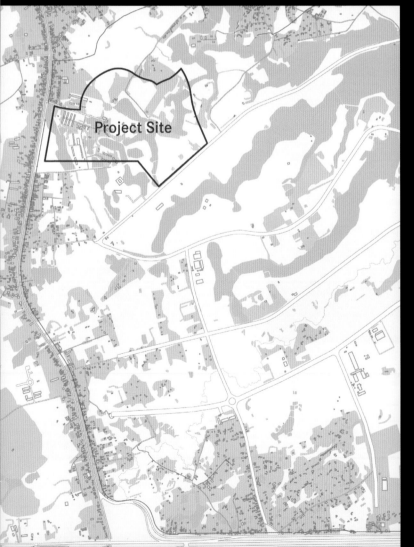

概况

业主：越南社会主义共和国（以越南教育培养部为代表，与亚洲发展银行合作）

项目规模：基地面积 65 公顷（143 000 平方米）

类型：开放型设计竞赛（根据亚洲发展银行的相关指导原则，设计招标由总承包商组织，各参赛者公开投标）（开放申请程序）

参赛者：6 组

竞赛预算：未规定

评委

建筑评委：Prof. Donald Bates, Architect, Melbourne; Dominique Lyon, Architect, Paris; Prof. em. Rodolfo Machado, Architect, Boston; Tobias Micke, Landscape Architect, Berlin; Dr. Tran Thanh Binh, Architect, Hanoi

专家评委：Prof. Bui Van Ga, Vice Minister, Deputy Minister of Ministry of Education and Training; Dr. Nguyen Van Ngu, Director Project Management Unit — University of Science and Technology Hanoi (PMU-USTH); Prof. Dr. Le Tran Binh, Vice Rector, University of Science and Technology Hanoi; Florence Kohler, Representative of France

"该项目是越南的标志性项目。河内科技大学由越南和法国合作共建,融合了两国的文化形态,象征着两国文化的深度交流。

教育、科研和技术设施应符合欧洲或国际标准,等同于美国普林斯顿大学标准,并具有法国品质。校园设计应该在体现上述原则的基础上反映越南的气候条件、历史文化、居民生活习惯。"

——引自《竞赛摘要》

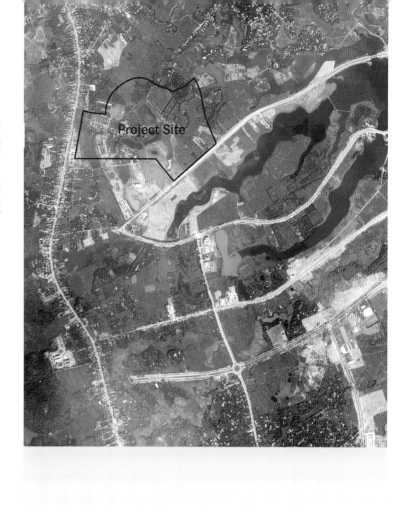

河内科技大学

与越南-德国大学相同,河内科技大学也是越南与合作伙伴国共同建立的四所"示范大学"之一,这四所大学作为高等教育机构应该符合国际标准。该大学建于 2009 年,法国为合作伙伴国。该项目合作由越南教育培训部和法国教育部牵头组织。

法国 40 所大学联盟是该项目的另一个战略合作伙伴。项目经费多半来自亚洲发展银行的贷款,亚洲发展银行的采购规则在项目中同样适用。

和乐高科技园区

河内科技大学坐落于和乐高科技园区,在河内西侧 37 千米。这里将成为由科研和教育机构组成的科学与技术中心。在越南,其地位相当于美国的硅谷或印度的班加罗尔。到目前为止,大约 65 公顷的土地为军事和农业用地,场地为丘陵地貌,有大面积地表水。用地西侧以 21 号国道为界,东南侧以公路 E 为界,东北侧以北 1 号公路为界。

竞赛任务

竞赛任务是通过组织竞赛来确定和乐高科技园区和河内科技大学校园总体规划的发展布局。校园规划分为两期:占地 25 公顷的建筑将于 2020 年前建成,供 5000 名学生使用;占地 40 公顷的建筑将于 2030 年前建成,供 15 000 名学生和 2200 名教职员工使用。该大学在建成后将包括教学楼及配套的教学设施(供 6 个院系使用)、图书馆、食堂、行政办公楼、学生及教职员工宿舍、运动场及配套的运动设施。

河内科技大学校园,河内

参赛者

1 一等奖 AS. Architecture Studio, Paris (F)
2 二等奖 Coelacanth and Associates (C+A), Nagoya (J)
3 三等奖 Auer + Weber + Assoziierte, Munich (D)
4 四等奖 AREP Ville, Paris (F)
5 五等奖 CPG Consultants, Singapore (SGP)
6 六等奖 D'Appolonia, Genoa (I)

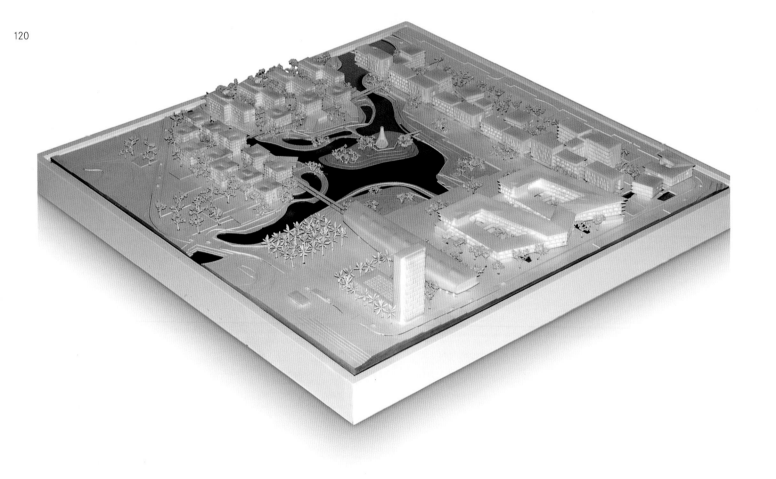

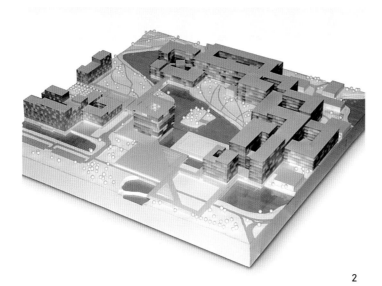

2

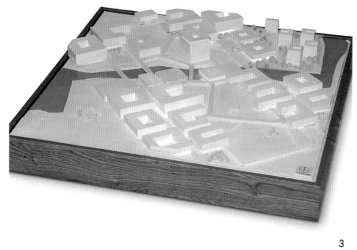

3

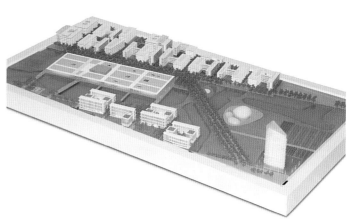

4

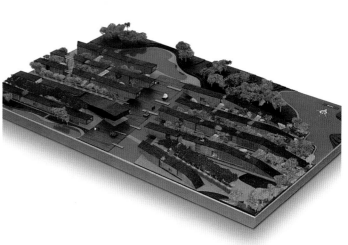

5

6

竞赛一等奖 获奖事务所

AS. Architecture Studio

Paris (F); with VHA Architects, Hanoi (VN); INGEROP International, Courbevoie (F); Project BASE, Paris (F)

方案设计者：René Henri Arnaud

    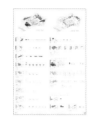  

VIEW OF CAMPUS PARK

VIEW OF SHARED FACILITIES DISTRICT

VIEW OF DORMITORIES

VIEW OF FACULTY

VIEW OF LEARNING RESOURCES CENTRE

UNIVERSITY OF SCIENCE AND TECHNOLOGY OF HANOI DEVELOPMENT (NEW MODEL UNIVERSITY) PROJECT

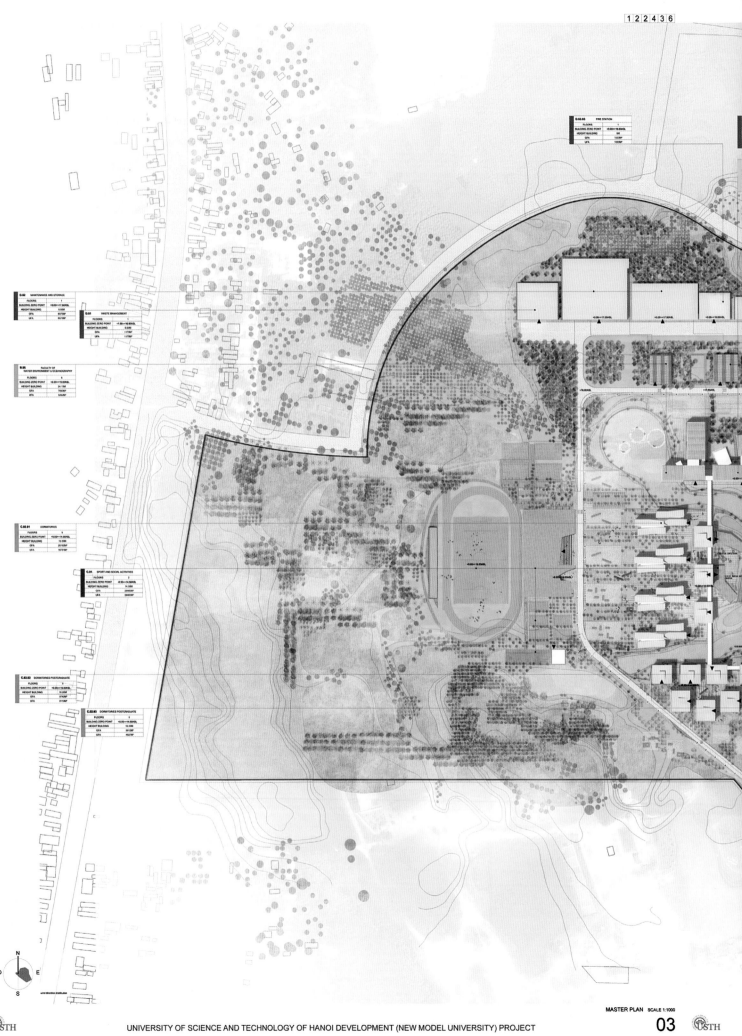

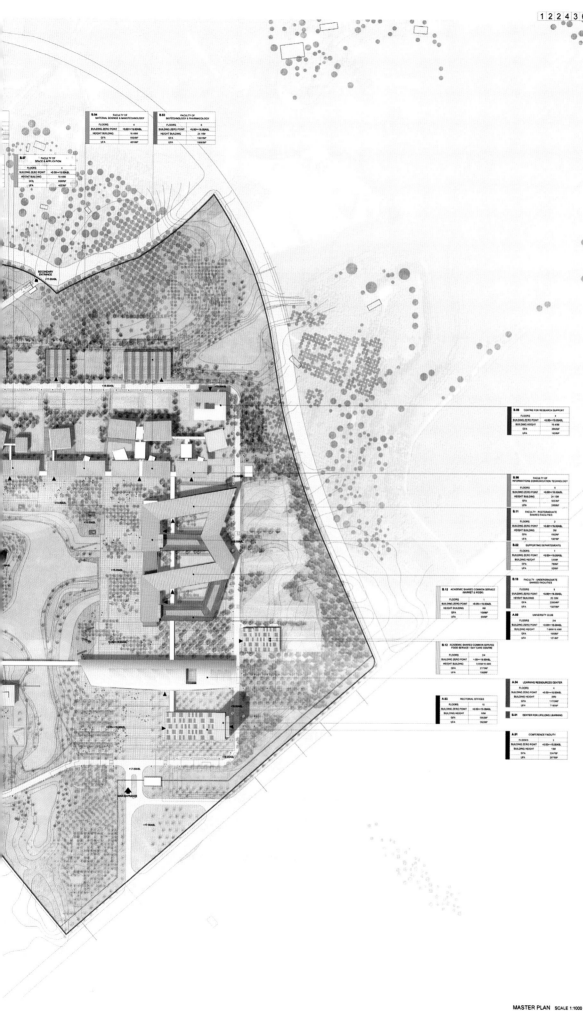

竞赛二等奖 获奖事务所

Coelacanth and Associates (C+A)

Nagoya (J); with Marina Stankovic Architekten, Berlin (D); PLACEMEDIA, Landscape Architects Collaborative, Tokyo (J); energydesign, Shanghai (CN); iproplan Planungsgesellschaft, Ho Chi Minh City (VN); Schlaich Bergermann und Partner, Stuttgart (D); und Langdon & Seah Vietnam, Hanoi (VN)

方案设计者：Yasuyuki Ito, Marina Stankovic, Tobias Jorzick, Shunsaku Miyagi, Prof. Dr. Dirk Schwede, Howard Stoneham, Gunnar Fassl, Marco R.disch, Knut G.ppert, Mark Andrew Olive
参与者：René Rabe, Holger Neumann, Ulrich Peters, Reinhard Kunze, Torsten Haugk, Jan Rautengarten, Christoph Wolter, Tran Van Tuan, Dao Kim Cuong, Tran Duc Loc, Pham Thanh Nguyen, Tran Giang Quynh, Nguyen Van Giang, Vuong Bich Thuy
顾问：iproplan Planungsgesellschaft, Chemnitz (D)

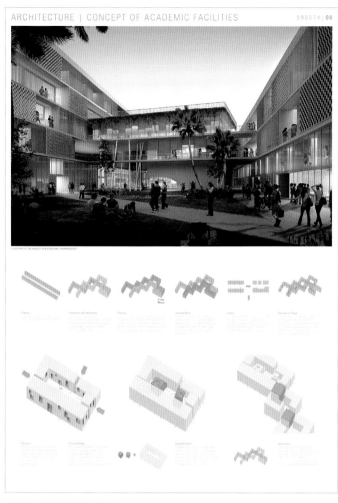
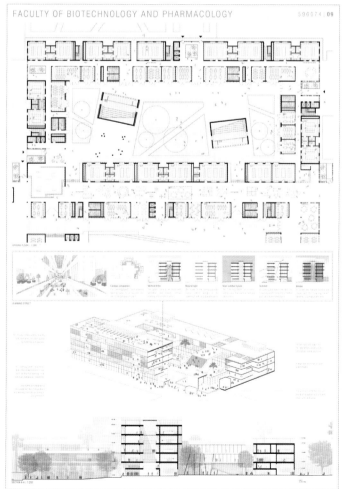

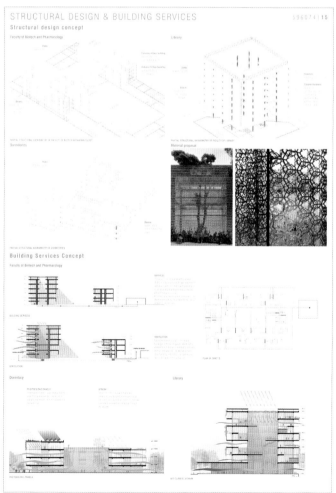

竞赛三等奖 获奖事务所

Auer + Weber + Assoziierte

Munich (D); with UBIK Architects, Hanoi (VN); Agence TER Paysagistes-Urbanistes, Paris (F); HBS, Hanoi (VN); and HL-PP Ingenieure International, Munich (D)

方案设计者: Stephan Suxdorf, Brice Belian, Henri Bava, Ngo Sy Lam, Klaus Peter
顾问: Consultancy Company for High-Rise Building Structure (VN); Tisseyre + Associés, Toulouse (F); Langdon Seah Vietnam, Hanoi/Ho Chi Minh City (VN)

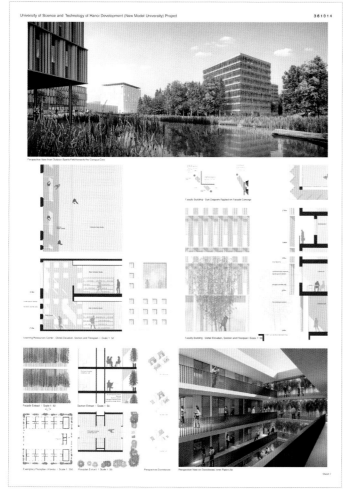
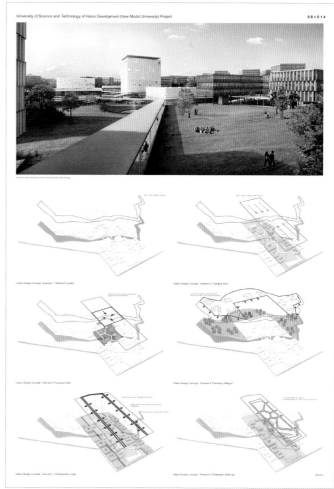

竞赛四等奖　获奖事务所

AREP Ville

Paris (F); with Brunet Saunier Architecture, Paris (F);
National General Construction Consulting, Ho Chi Minh City (VN)

方案设计者：Gael Desveaux, Etienne Tricaud, Jerome Brunet, Daniel Claris, Christophe Chevallier, Chau Nguyen, Thierry Noblesse, Andreas Alexopoulos, Thinh Ho, Wim Boydens, Y Nguyen, Thach Tran
顾问：Florian Ligier, Kien Tran, Duc Doan, Thai Nguyen

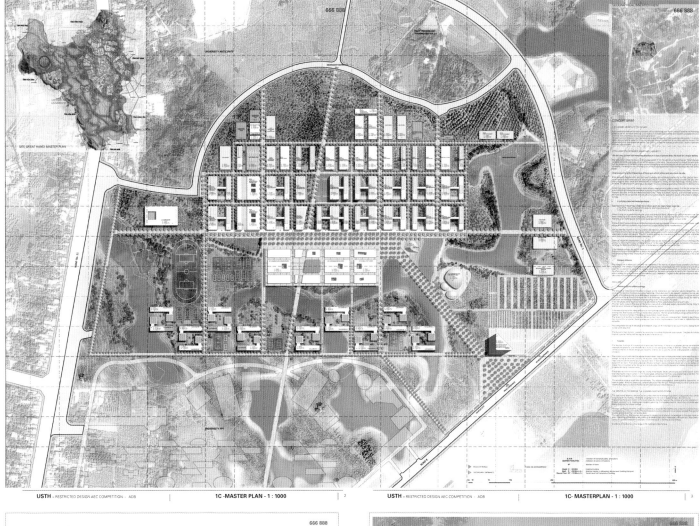
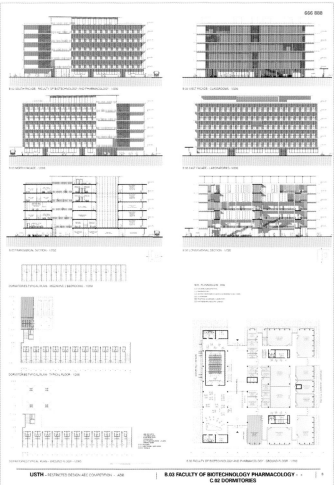

AREP Ville

河内科技大学校园，河内

竞赛五等奖　获奖事务所

CPG Consultants

Singapore (SGP); with Hanoi Design Construction, Hanoi (VN);
Studio Milou Architecture, Paris (F)

方案设计者： Kuan Chee Yung, Jimmy Liu Wing Tim, Ng Kim Leong,
Le Viet Son, Jean Francois Milou, Florence Soulier, Ng Kim Leong
顾问： CPG Vietnam (VN); CCW Associates, Singapore (SGP)

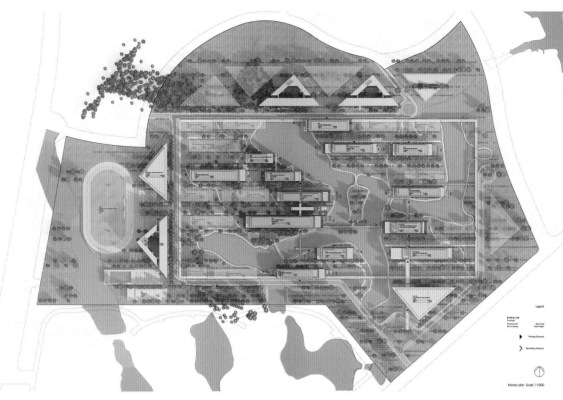

Concept Statement

The project offers an elegant ensemble of structures and landscaping designed to meet all of the brief's functional requirements while respecting the natural beauty of the site and the organic expansion of the campus in the future. Coherent and discrete design unity characterizes the project. Priority is given to elegant forms and a unity of colors and materials against the backdrop of abundant vegetation and outdoor spaces. The layout of the design echoes the appreciation of symmetry that often typifies French architectural traditions, and creates spaces conducive to a sense of order, meditation and reflection. More importantly, the Master plan is influenced by the context of the Hanoi Natural and Cultural landscape:

- The tropical water-way farmland nature of the site that seeks to integrate the village environment with existing seasonal water-bodies and retain the surrounding lush forest.
- The cultural academic tradition of Hanoi conceived as a series of courtyards, lakes, colonnade walkways and gates housing linear "temples of learning" & court of contemplation & discussion.

The coherence and apparent simplicity of the design have been achieved only through a rigorous and complex study of the brief in relation to the site and its surrounding urban environment.

3 - 4. A changing garden for education
This garden is composed as a fabric of trees, fences and water pounds, changing with the seasons and the water level on site, providing a mix of wet gardens and dry landscape

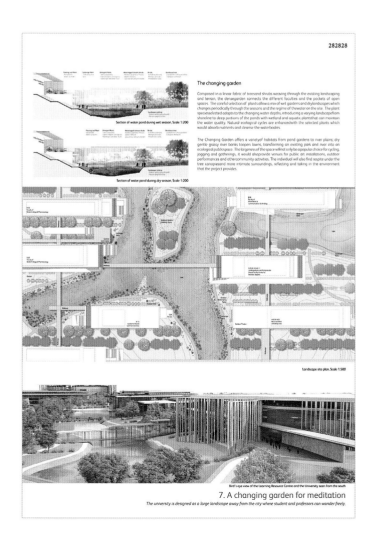

7. A changing garden for meditation
The university is designed as a large landscape away from the city where student and professors can wander freely.

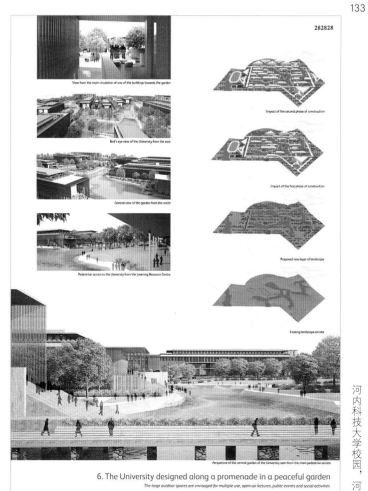

6. The University designed along a promenade in a peaceful garden
The large outdoor spaces are envisaged for multiple use, open-air lectures, public events and social activities.

竞赛六等奖　获奖事务所

D'Appolonia

Genua (I); with IDEAS (ESTUDIO); Rome (I); Caravaggi Lucina Architect, Rome (I); IDEAS, Milan (I)

方案设计者：Roberto Carpaneto, Prof. Andrea Del Grosso, Roberto Carpaneto, Guendalina Salimei, Mavilio Stefano, Ferrari Rocco Carlo, Tung Le, Vecchi Antonio, Lucina Caravaggi, Cristina Imbroglini, Giancarlo Fantilli, Rosario Pavia, Beatrice Majone, Emanuel Habib, Alessandro Balbo

参与者：Paola Silva, Paolo Basso, Andrea Tomarchio, Fabio Figini, Maurizio Iannolo, Lorenzo Ruffini, Alessandro Aliotta, Valentina Morelli, Stefania Banchero, Riccardo Viviano, Enrico Grosso, Fabrizio Tavaroli, Michele Mililli, Giovanna Lo Re, Isabella Severi, Laura Zevi, Manuel Lentini, Riccardo Pagnanini, Fabiana Marchesi, Roberta Romiti, Marina Simonetti

自由职业者：Claudia Bodesmo, Bonino Ruggero, Gianni Carletti, Fabio Ferone, Giulia Giampiccolo, Andrea Laganà, Elena Immè, Luisella Pergolesi, Giulio Tonelli, Andrea Paduano, Mario Ferrari, Anna Peronace, Diego Bianchi

顾问：Italian Design Engineering Architecture Solutions (VN); EPEL Construction Engineering Company (VN)

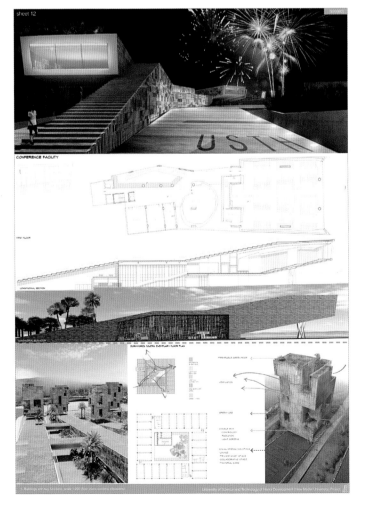

蒂森克虏伯房地产有限公司办公楼

德国 柏林

2011

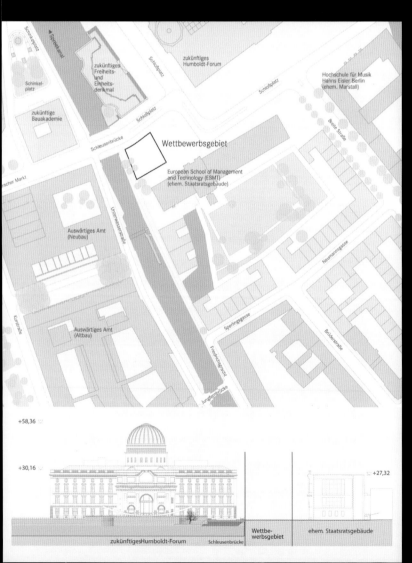

概况

业主：蒂森克虏伯房地产有限公司
项目规模：基地面积约 5000 平方米
类型：双阶段开放型设计竞赛（开放申请程序）
参赛者：第一阶段 30 组，第二阶段 7 组
竞赛预算：154 000 欧元（奖金 56 000 欧元，第二阶段每组参赛者的费用 16 000 欧元）

评委

Mels Crouwel, Architect, Amsterdam; Prof. Dietrich Fink, Architect, Munich; Prof. Ulrike Lauber, Architect, Berlin/Munich; Prof. Manuel Scholl, Architect, Zurich; Dr. Jürgen Claassen, Member of the Board, ThyssenKrupp AG, Essen; Ralph Labonte, Member of the Board, ThyssenKrupp AG, Essen; Regula Lüscher, Architect, Senate Building Director/Permanent Secretary, Senate Department for Urban Development and the Environment Berlin; Dr. Martin Grimm, Chief Executive Officer, ThyssenKrupp Real Estate GmbH, Essen; Ephraim Gothe, District Councillor for Urban Development, District Berlin-Mitte

"竞赛任务是设计位于德国柏林的蒂森克虏伯房地产有限公司办公楼（一座集科研、艺术、办公、居住于一体的综合体），并在丰富多彩的环境中设置多个代表机构。项目面临的挑战在于场地的矛盾和建筑的功能：既要参考历史又要面向未来。"

——引自《竞赛摘要》

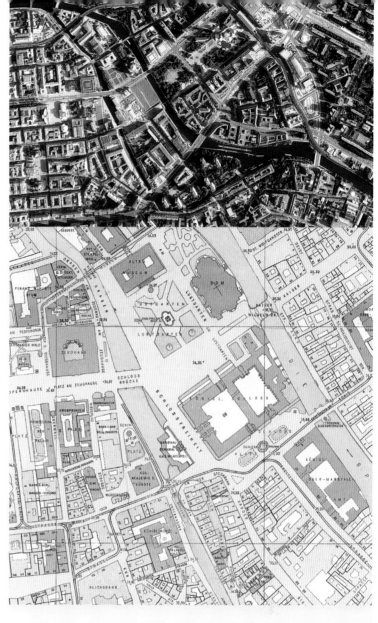

蒂森克虏伯房地产有限公司办公楼

蒂森克虏伯房地产有限公司办公楼地处德国柏林的突出位置，是可激发灵感的会议研讨空间，是技术创新和技术服务的平台，也是举办大型活动和展览的场所。项目旨在使该大楼在大的公共尺度和小的私密性方面与柏林其他著名建筑形成映衬和呼应。因此，既有的公司理念应该更新。蒂森克虏伯房地产有限公司作为德国企业文化的全球运营商，其办公楼应该更关注技术创新。

柏林皇宫广场

700平方米的竞赛场地位于德国柏林的皇宫广场，这里形成于20世纪90年代，是柏林的历史文化中心。蒂森克虏伯房地产有限公司办公楼地处皇宫广场西南边缘，是城市结构完善计划的一部分。该大楼在未来将成为洪堡论坛的所在地。该地段具有历史矛盾性：这里曾经是柏林世界主义思潮极其活跃的中心，之后又是德意志民主共和国的政治中心。在未来，洪堡广场、洪堡大学、博物馆和其他机构（例如，申克尔建筑学院、欧洲管理与技术学校）、德国国家理事会办事处将赋予该区域动态特征。然而，随着竞赛的开展，德国柏林参议院决定禁止该大楼建于此地。

竞赛任务

该项目面积约2800平方米，规划区域包括办公室、会议室以及供接待、研讨之用的多功能用房。材料美学、色彩搭配和光线概念应该符合该大楼地处创新之地的得天独厚的地理位置。此外，方案设计应该充分考虑基地历史文化遗址保护的问题，比如，工程建设不应对场地中的历史墙体和紧临Staatsratsgebäude区域的历史建筑造成破坏，而应使它们与新建大楼形成一定的呼应。

蒂森克虏伯房地产有限公司办公楼，柏林

第一阶段参赛者

第二阶段入围者

1 Thomas Müller Ivan Reimann Architekten, Berlin (D) **2** Hascher Jehle Architektur, Berlin (D) **3** Grüntuch Ernst Architekten, Berlin (D) **4** JSWD Architekten, Cologne (D), and Atelier d'architecture Chaix & Morel et Associés, Paris (F) **5** Schultes Frank Architekten, Berlin (D) **6** Schweger Associated Architects, Hamburg (D) **7** Kaspar Kraemer Architekten, Cologne (D)

其他参赛者

8 Barkow Leibinger Architekten, Berlin (D) **9** Nieto Sobejano Arquitectos, Berlin (D) **10** Wingårdh Arkitektkontor, Gothenburg (S) **11** Kusus + Kusus Architekten, Berlin (D) **12** Claus en Kaan Architecten, Rotterdam (NL) **13** gmp Architekten, Berlin (D) **14** Gatermann + Schossig Bauplanungsgesellschaft, Cologne (D) **15** Staab Architekten, Berlin (D) **16** Max Dudler Architekten, Berlin (D) **17** Stephan Braunfels Architekten, Berlin (D) **18** meck architekten, Prof. A. Meck Architekturbüro, Munich (D) **19** Peter Kulka Architektur, Cologne (D) **20** Ortner & Ortner Baukunst, Berlin (D) **21** kadawittfeldarchitektur, Aachen (D) **22** Auer + Weber + Assoziierte, Stuttgart (D) **23** Steidle Architekten, Munich (D) **24** Architekturbüro Professor Wolfgang Kergaßner, Ostfildern (D) **25** Degelo Architekten, Basel (CH) **26** Oskar Leo Kaufmann | Albert Rüf, Dornbirn (A) **27** Mila/Jakob Tigges, Berlin (D)

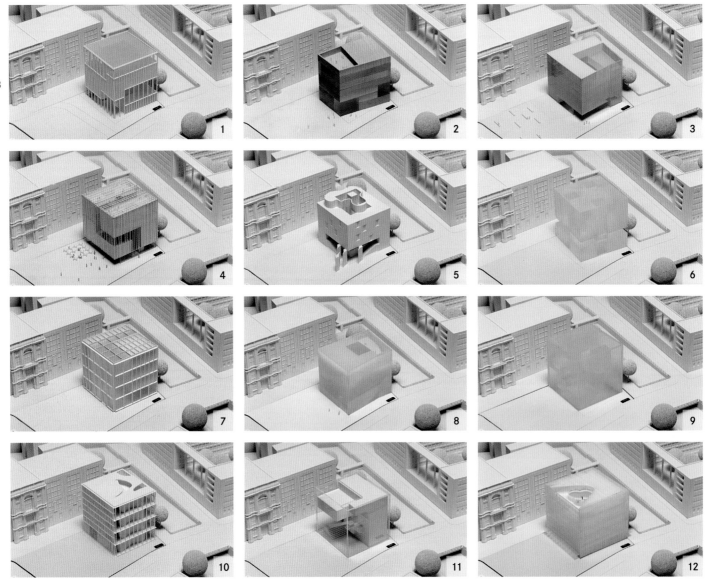

第一阶段参赛者

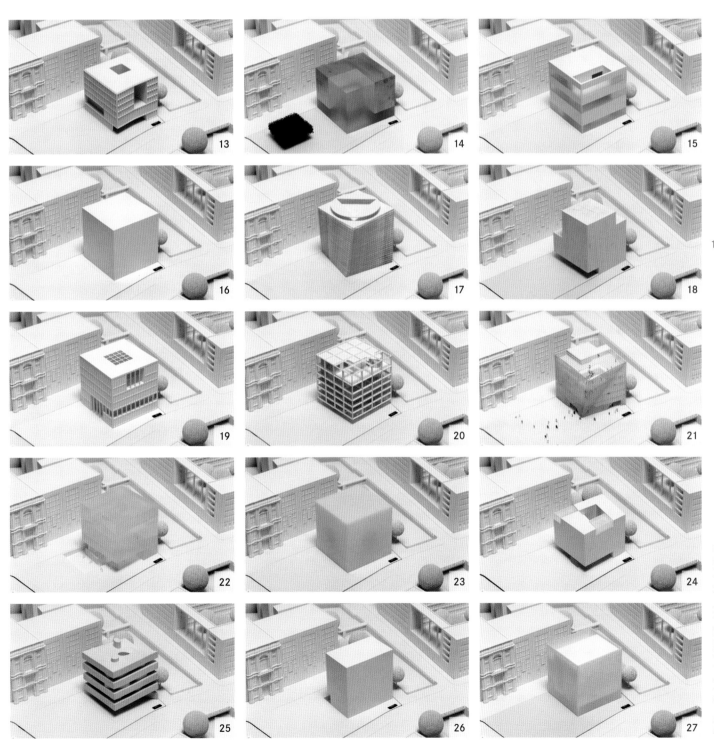

蒂森克虏伯房地产有限公司 办公楼，柏林

第二阶段参赛者

获奖者
1 一等奖　Schweger & Partner Architekten, Hamburg (D)
2 二等奖　JSWD Architekten, Cologne (D), and Atelier d'architecture Chaix & Morel et Associés, Paris (F)
3 二等奖　Grüntuch Ernst Architekten, Berlin (D)
4 二等奖　Kaspar Kraemer Architekten, Cologne (D)

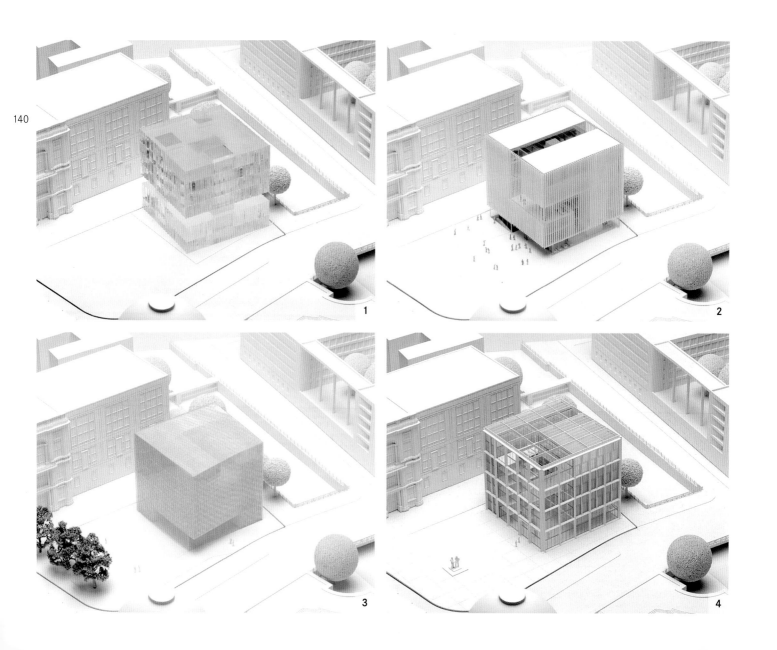

第二轮

5 Thomas Müller Ivan Reimann Architekten, Berlin (D)
6 Schultes Frank Architekten, Berlin (D)
7 Hascher Jehle Architektur, Berlin (D)

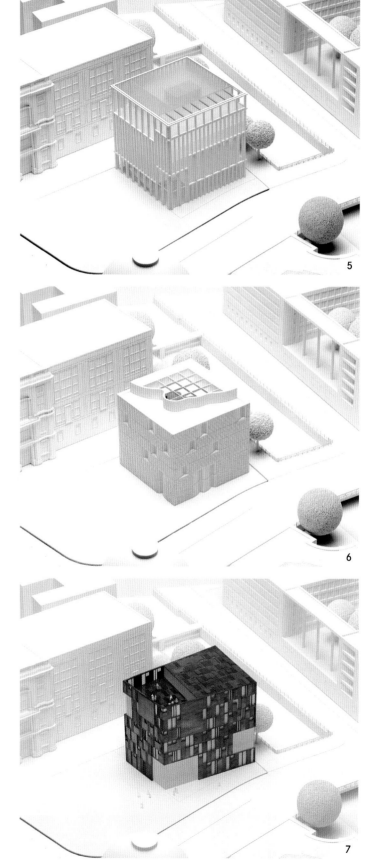

竞赛一等奖　获奖事务所

Schweger & Partner Architekten

Hamburg (D)

方案设计者：Peter P. Schweger, Mark Schüler, Jens-Peter Frahm
参与者：Ralf Hawer
顾问：Werner Sobek Stuttgart (D); Ridder und Meyn Ingenieurgesellschaft, Berlin (D); Peter Andres, Hamburg (D)

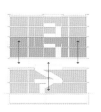
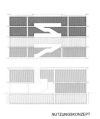

TECHNIK / LOGISTIK   UG
EINGANG/CAFE   EG
SONDERNUTZUNG   1.OG
BELVEDERE/TERRASSE   2.OG

ZIRKULATIONSKONZEPT

TK KONFERENZ   3.OG
TK BÜRO   4.OG
PARTNER BÜRO   5.OG
PARTNER BÜRO   6.OG

NUTZUNGSKONZEPT

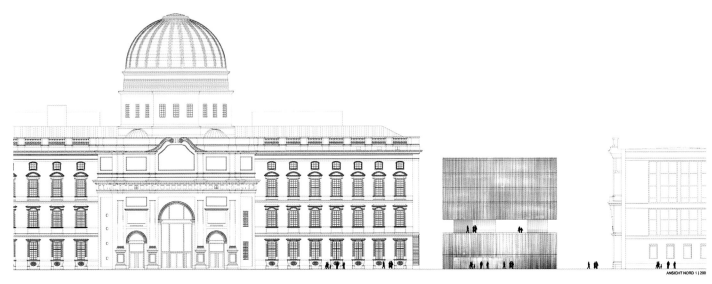

ANSICHT NORD 1 | 200

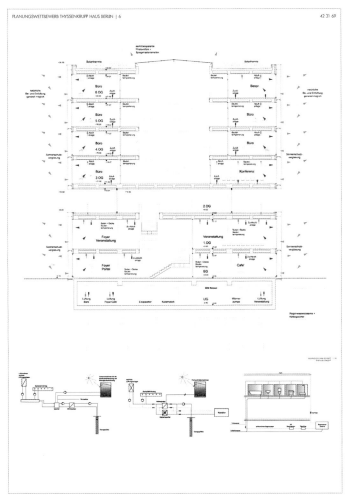
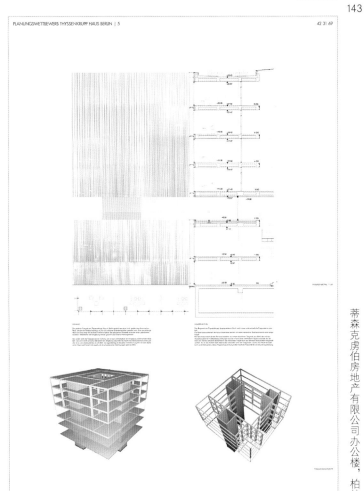

竞赛二等奖 获奖事务所

JSWD Architekten

Cologne (D); with Atelier d'architecture Chaix & Morel et Associés, Paris (F)

方案设计者：Frederik Jaspert, Walter Grasmug
参与者: Carolin Amann, Fabien Barthelemey, Svea Gerland, Jan Horst, Til J.ger, Linh Le, Christian Mammel, Martin Oehme, Bogna Przybylska, Cecile Riviere, Yohanna Vogt, Maximilian Wetzig
自由职业者：Eddie Young
顾问：WSGreenTechnologies, Stuttgart (D); Werner Sobek Stuttgart (D); BFT Cognos, Aachen (D); KLA kiparlandschaftsarchitekten, Duisburg (D)

竞赛二等奖　获奖事务所

Grüntuch Ernst Architekten

Berlin (D)

方案设计者： Armand Grüntuch
参与者： Stefan Schenk, Dominik Queck, Carolin D.pfer, Thiele Nickau
顾问： Prof. Klaus Daniels, HL-Technik, Munich (D); Schlaich Bergermann Partner, Stuttgart (D), Maik Schlaich

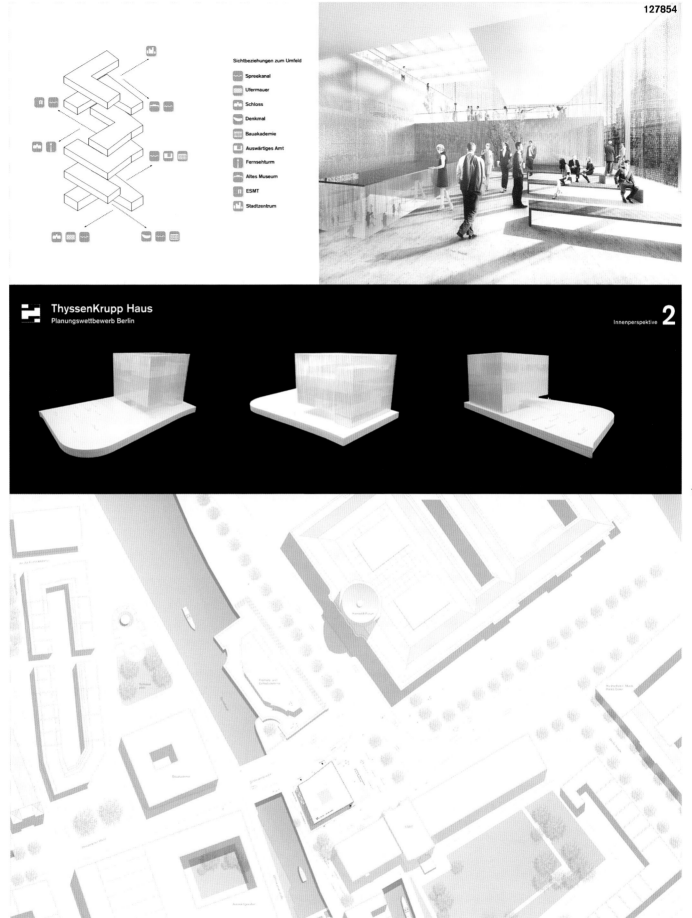

竞赛二等奖　获奖事务所

Kaspar Kraemer Architekten

Cologne (D)

方案设计者：Kaspar Kraemer
参与者：Marcel Jansen, Daniel B.ger, Nina Schilling
自由职业者：Hans-Günter Lübben
顾问：Pirlet + Partner, Cologne (D); BSCON Brandschutzconsult, Essen (D); Club L94 Landschaftsarchitekten, Cologne (D), Pfeil + Koch Ingenieurgesellschaft, Cologne (D)

ThyssenKrupp Haus Berlin

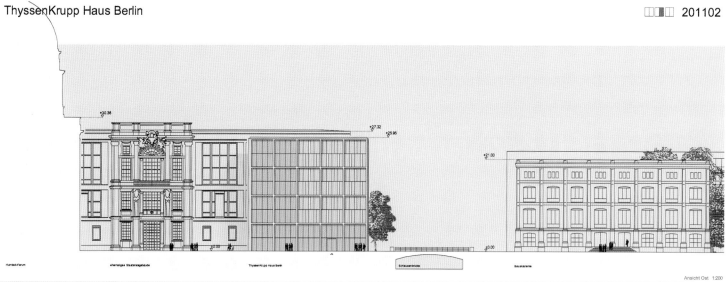

Ansicht Ost 1:200

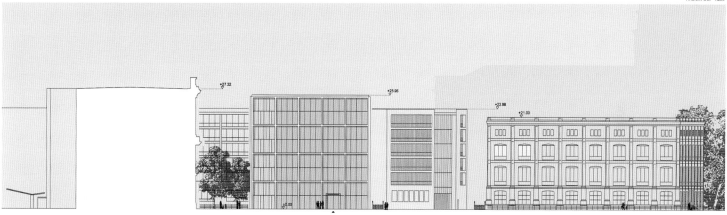

Ansicht Nord 1:200

Raumprogramm ThyssenKrupp

Öffentlichkeit Erdgeschoss

Funktion

Die Eingangsebene versammelt in der ‚Stadtloggia' Öffentlichkeit und Besucher der Repräsentanz. Die Pforte übernimmt Information, Aufsicht und Zugangskontrolle. Der Ausstellungsbereich und das Café im Erdgeschoss sind offen miteinander verbunden, können aber bei Bedarf durch eine Faltwand getrennt werden. Zur ESMT im Süden sind in einer Funktionsspange über alle Geschosse die Nebenfunktionen orientiert. Über die gläsernen Aufzüge erreichen die Besucher und die Partner in den oberen Nutzungsebenen die Büros der Hauptstadtrepräsentanz von ThyssenKrupp.

Im 1. Obergeschoss ist eine optionale Ausstellungsebene als Galerie konzipiert. Im 2. und 3. OG sind die Partner-Ebenen angeordnet, im 4. OG befindet sich das Berlinbüro von ThyssenKrupp, im 5. OG befindet sich die Konferenzzone, im 6. OG der große Saal, als Sondernutzung mit Dachterrasse.

Die Grundrisse sind flexibel und offen gestaltet und passen sich der jeweiligen Anforderungen an Aufzüge und Treppen verbinden die Ebenen und gewähren Kommunikation und Zusammenschluss. Über einen Vorbereich in der Stadtloggia erschließen sich die Büro-, Konferenz- und Sondernutzungsebenen.

Die flexible Grundstruktur des Hauses mit Funktionsstücken im Süden und frei bespielbaren Flächen auf allen Ebenen ermöglicht die flexible Anordnung der Nutzung je nach Bedarf. Alle Nutzungen sind nach Westen, Norden und Osten orientiert und garantieren den Ausblick auf die historische Mitte der Hauptstadt mit Auswärtigem Amt, Bauakademie, Kommandantur, Zeughaus, Lustgarten und Altes Museum sowie Humboldt-Forum mit Schlossplatz, Marstall und ehemaligem Staatsratsgebäude.

Außenbereich

Vom Schlossplatz mit Vorfahrt im Osten bis zum Café zum Kupfergraben wird das ThyssenKrupp Haus im Westen und die Platzfläche samt der Eingangsebene mit einem einheitlichen Steinbelag gestaltet, der das Konstruktionsraster des ThyssenKrupp-Hauses aufnimmt und dieses mit dem historischen Plattenmuster des ehemaligen Staatsratsgebäudes verzahnt. Im Schnittpunkt der Achslinien der jeweiligen Engänge von ThyssenKrupp-Haus und ESMT könnte eine Figurengruppe Aufstellung finden, die analog zum Goethe/Schiller-Denkmal in Weimar Alfred Krupp und August Bebel in Beziehung zueinander setzt um die Öffentlichkeit zum Betreten der Hauptstadtrepräsentanz einzuladen. Zwischen Neubau und ESMT werden die Behinderten- und Fahrradstellplätze am Sockelgeschoss der ESMT angeordnet. Aus dem städtischen Umfeld wird so das ThyssenKrupp-Gebäude als nach allen Seiten offenes und zugängliches Haus erfahren, ohne den Anspruch an Sicherheit und notwendige Kontrolle einzuschränken.

Blick vom Auswärtigen Amt

Schnitt A-A 1:200

Schnitt B-B 1:200

50 Hertz 公司总部大楼

德国 柏林

2012—2013

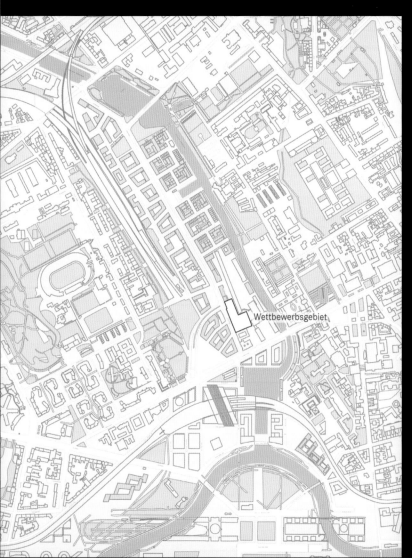

概况
业主：50 Hertz 传输股份有限公司
项目规模：基地面积约 18 000 ~ 25 000 平方米
类型：开放型设计竞赛（开放申请程序）
参赛者：18 组
竞赛预算：220 000 欧元（奖金 101 000 欧元，荣誉奖 11 000 欧元，每组参赛者的费用 6000 欧元）

评委
建筑评委：Prof. Markus Allmann, Architect, Munich; Prof. Stefan Behnisch, Architect, Stuttgart; Prof. Almut Grüntuch-Ernst, Architect, Berlin; Stefan Plesser, Architect Brunswick; Prof. Kirsten Schemel, Architect, Berlin; Prof. Gernot Schulz, Architect, Cologne; Prof. Jürgen Weidinger, Landscape Architect, Berlin
专家评委：Boris Schucht, Chief Executive Officer, 50 Hertz Transmission GmbH; Hans-Jörg Dorny, Personnel Director, 50 Hertz Transmission GmbH; Dr. Lutz Pscherer, Chairman of the Works Council, 50 Hertz Transmission GmbH; Regula Lüscher, Senate Building Director/Permanent Secretary, Senate Department for Urban Development and the Environment Berlin; Carsten Spallek, District Councillor for Urban Development, Construction, Economics and Regulation, District Berlin-Mitte; Henrik Thomsen, Head of CA Immo Berlin, CA Immobilien Anlagen GmbH

"位于德国柏林的 50 Hertz 公司总部大楼应充分体现 50 Hertz 公司的四个核心理念：开放、透明、合作、交流。"
——引自《竞赛摘要》

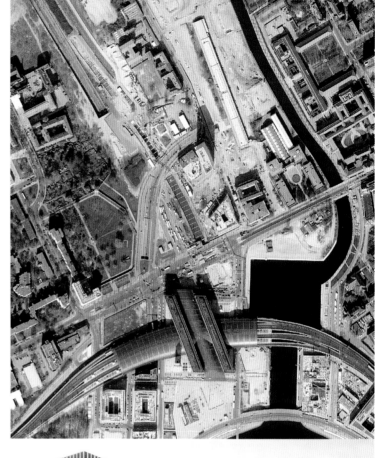

50 Hertz 公司

网络运营商 50 Hertz 公司的主要业务为商业网络供电、电力输送和安全能源供给，服务于德国北部和东部的约 1800 万人口。50 Hertz 公司拥有大约 800 名员工，为中型公司，由德国联邦网络局管理。50 Hertz 公司总部大楼的竞赛任务是制订一个具有创新性和经济性的建筑方案，营造适宜的空间环境，促进员工之间的业务和文化交流，体现"开放、透明、合作、交流"的公司理念。

海德路 - 欧罗巴城市规划

8145 平方米的开发用地是海德路 - 欧罗巴城市规划的一部分，超过 40 公顷的地块沿着柏林中心车站以北的柏林 - 施潘道运河展开。新的开发区将连接北部 Wedding 区、中心车站和南部政府区。为了顺利地推行城市规划，参议院与柏林米特区办公室决定以中心车站为中心制订区域规划——根据各城市区域的不同特点，建立不同的规划框架，创建办公、居住、购物、文化休闲等不同的建筑类型，满足不同目标群体和用户的需求。

竞赛任务

项目重点在于将基地面积由 18 000 平方米拓展至 25 000 平方米，并将人口容量提升至 580 人。办公室、会议室、日托中心和传输控制中心均处于该建筑的核心部分。传输控制中心是一个高度安全的控制室，用于监控和管理电力传输网络。

参赛者

获奖者
1　一等奖　LOVE architecture and urbanism, Graz (A)
2　二等奖　Henning Larsen Architects, Copenhagen (DK), Munich (D)
3　三等奖　alexa zahn architekten with swap architekten, Vienna (A)
4　四等奖　Thomas Müller Ivan Reimann Architekten, Berlin (D)
5　荣誉奖　KSP Jürgen Engel Architekten, Frankfurt am Main (D)
6　荣誉奖　wwa - w.hr heugenhauser architekten, Vienna (A)

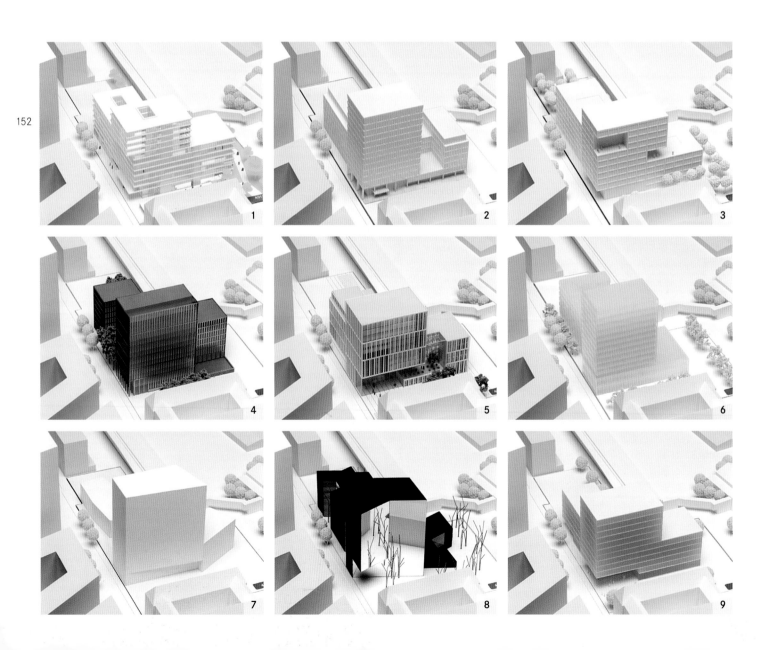

第二轮

- 7 Dietrich | Untertrifaller Architekten, Vienna (A)
- 8 NO.MAD Arquitectos, Madrid (E)
- 9 h.s.d. architekten, Lemgo (D)
- 10 Kleihues + Kleihues Architekten, Berlin (D)
- 11 Max Dudler Architekten, Berlin (D)
- 12 Sauerbruch Hutton Architekten, Berlin (D)
- 13 Goetz Hootz Castorph Planungsgesellschaft, Munich (D)
- 14 Molestina Architekten, Cologne (D), with FSWLA, Düsseldorf (D)
- 15 Atelier Thomas Pucher, Graz (A)

第一轮

- 16 Hadi Teherani Architects, Hamburg (D)
- 17 Anton Meyer Architekt, Dachau (D)
- 18 Lothar Jeromin Architekt, Essen (D)

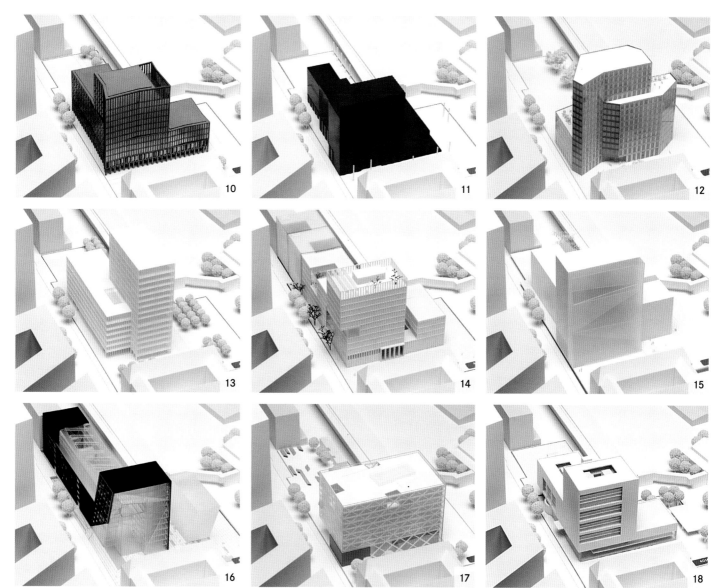

竞赛一等奖　获奖事务所

LOVE architecture and urbanism

Graz (A)

方案设计者：Mark Jenewein, Herwig Kleinhapel, Bernhard Sch.nherr
参与者：Julius Popa, Robin Bohman
自由职业者：Eva Sollgruber, Ines Escauriaza Otazua, Adrian Armasescu-Fusa, Vit Music
顾问：Hartmut Sommer, Hamburg (D); Dr. Roland Müller, Stockerau (D); Wetzel von Seth, Berlin (D); Bartenbach Lichtlabor, Aldrans (A)

Wettbewerb - 50hertz Netzquatier

704365

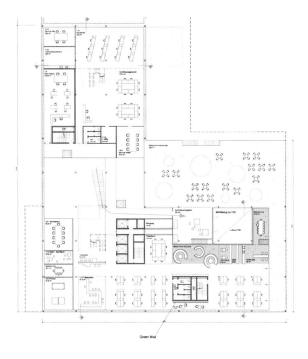

Grundriss 1.OG 1:200

Grundriss 2.OG 1:200

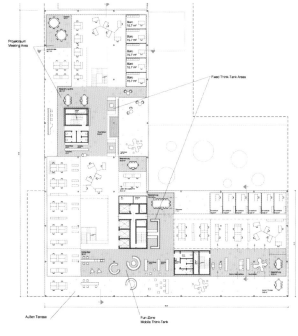

Grundriss 4.OG 1:200

Grundriss UG 1:500

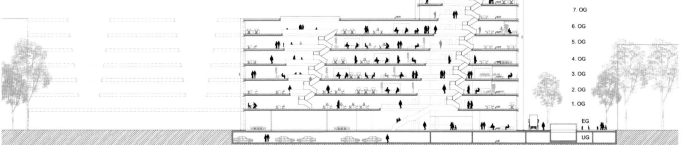

Schnitt A 1:200

竞赛二等奖　获奖事务所

Henning Larsen Architects

Copenhagen (DK), Munich (D)

方案设计者：Louis Becker, Werner Frosch
参与者：Juliane Demel, Daniel Baumann, Mikala Holme Samsoe, René Anderson, Eileen Dorer, Omar Dabaan
顾问：Happold Ingenieurbüro, Berlin (D); El:ch Landschaftsarchitekten, Berlin (D)

Vernetzung: Räumliche Flexibilität

Schnitt A-A 1:200

BMW FIZ Future

德国 慕尼黑

2013—2014

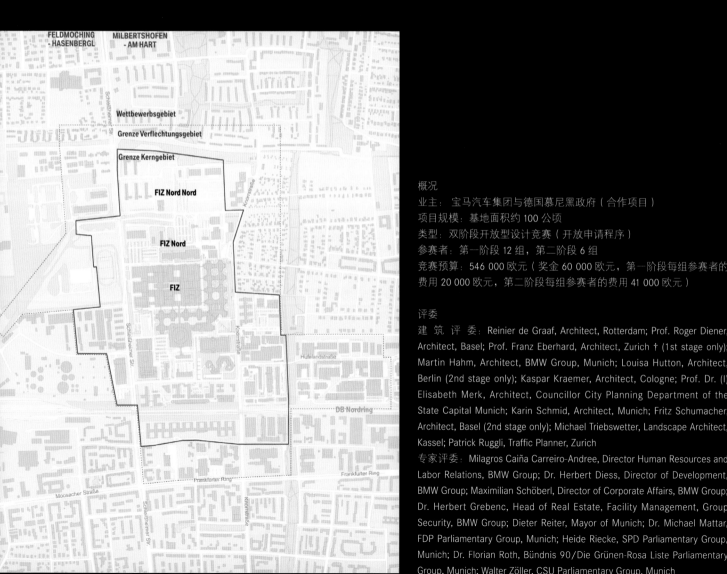

概况

业主：宝马汽车集团与德国慕尼黑政府（合作项目）
项目规模：基地面积约 100 公顷
类型：双阶段开放型设计竞赛（开放申请程序）
参赛者：第一阶段 12 组，第二阶段 6 组
竞赛预算：546 000 欧元（奖金 60 000 欧元，第一阶段每组参赛者的费用 20 000 欧元，第二阶段每组参赛者的费用 41 000 欧元）

评委

建 筑 评 委：Reinier de Graaf, Architect, Rotterdam; Prof. Roger Diener, Architect, Basel; Prof. Franz Eberhard, Architect, Zurich † (1st stage only); Martin Hahm, Architect, BMW Group, Munich; Louisa Hutton, Architect, Berlin (2nd stage only); Kaspar Kraemer, Architect, Cologne; Prof. Dr. (I) Elisabeth Merk, Architect, Councillor City Planning Department of the State Capital Munich; Karin Schmid, Architect, Munich; Fritz Schumacher, Architect, Basel (2nd stage only); Michael Triebswetter, Landscape Architect, Kassel; Patrick Ruggli, Traffic Planner, Zurich

专家评委：Milagros Caiña Carreiro-Andree, Director Human Resources and Labor Relations, BMW Group; Dr. Herbert Diess, Director of Development, BMW Group; Maximilian Schöberl, Director of Corporate Affairs, BMW Group; Dr. Herbert Grebenc, Head of Real Estate, Facility Management, Group Security, BMW Group; Dieter Reiter, Mayor of Munich; Dr. Michael Mattar, FDP Parliamentary Group, Munich; Heide Riecke, SPD Parliamentary Group, Munich; Dr. Florian Roth, Bündnis 90/Die Grünen-Rosa Liste Parliamentary Group, Munich; Walter Zöller, CSU Parliamentary Group, Munich

"竞赛理念应该围绕技术创新和提高城市生活质量。根据现有建筑结构和城市布局的基本特征，提出创新的、前瞻性的建筑设计和城市规划提案。这一实施过程，究其实质，是一番明智、巧妙、合理的"转译"，即将汽车技术研发中心转变成具有美感与吸引力、满足不同分区的功能需求且鼓舞人心的工作与生活之所。"

——引自《竞赛摘要》

宝马集团研究与创新中心
宝马集团研究与创新中心被认为是世界上最现代化的大型汽车技术研发中心之一，是宝马集团研究和开发业务的核心。从长远考虑，宝马公司力求巩固并扩大用地规模（目前可容纳 26 000 名员工），计划拓展约 100 万平方米的基地面积并优化其空间功能结构，进而为供应商及合作伙伴公司提供大约 5000 个工作岗位。为此，集团北部约 20 公顷的用地及现有设施被收购。该中心位于慕尼黑北部，靠近环路弗兰克 - 福特地段；其邻近宝马集团总部和主要生产设施，地理位置具有战略意义；同时，这里也是慕尼黑北部城市结构的重要组成部分。

竞赛任务
竞赛任务是为宝马集团研究与创新中心及周边地区制订一个中长期发展和结构调整的跨学科方案设计，旨在促进宝马集团、慕尼黑市政府以及市民和其他业主之间的交流与合作。总体规划应充分整合相关要素，以保持区域未来的可持续发展。除了中长期发展和结构调整，在2025—2050 年间应考虑绿地规划交通动线及机动车停靠对现有空间布局及未来布局调整的影响。到 2025 年，约增加 50 万平方米的基地面积。基于规模、地理位置和未来发展的重要性，竞赛组委会设定了100 公顷超出该中心边界的用地范围，分为初级核心区和次级互联区。

程序
为了应对复杂的竞赛任务，参赛者需要协调各方利益，并与各方相关利益者合作。竞赛通常以"城市会议"或"邻里对话"的形式进行，并通过跨学科与"合作"的程序得以实施。

BMW FIZ Future，慕尼黑

第一阶段参赛者

第二阶段入围者

1. AllesWirdGut Architektur, Vienna (A), with club L94 Landschaftsarchitekten, Cologne (D)
2. ernst niklaus fausch architekten, Zurich (CH), with Müller Illien Landschaftsarchitekten, Zurich (CH), and Rapp Infra, Basel (CH)
3. gmp International, Hamburg (D), with ST raum a. Landschaftsarchitekten, Berlin (D)
4. West 8 Urban Design & Landscape Architecture with Atelier Kempe Thill - Architects and Planners, Rotterdam (NL)
5. Henn Architekten, Munich (D), with Topotek 1, Berlin (D)
6. AS&P Albert Speer & Partner, Frankfurt am Main (D), with ver.de Landschaftsarchitektur, Freising (D)

其他参赛者

7. Clive Wilkinson Architects, Culver City, with Stoss Landscape Urbanism, Boston (USA)
8. DP Architects, Singapore (SGP)
9. KAAN Architekten, Munich (D), with OKRA Landschapsarchitecten, Utrecht (NL)
10. BIG - Bjarke Ingels Group, Copenhagen (D), with t17 Landschaftsarchitekten, Munich (D)
11. agps architecture with Vogt Landschaftsarchitekten, Zurich (CH)
12. White Arkitekter, Stockholm (S)

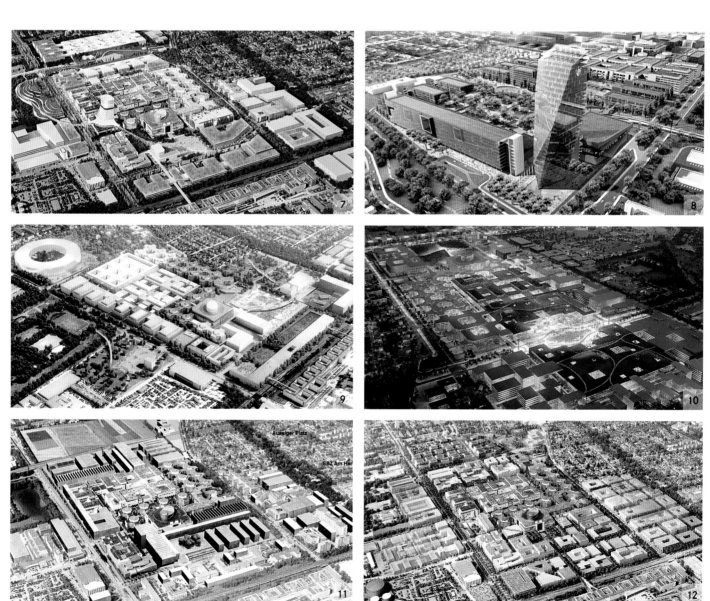

第二阶段参赛者

获奖者
1 一等奖 Henn Architekten, Munich (D), with Topotek 1, Berlin (D)
2 二等奖 ernst niklaus fausch architekten, Zurich (CH), with Müller Illien Landschaftsarchitekten, Zurich (CH), and Rapp Infra, Basel (CH)
3 三等奖 West 8 Urban design & Landscape Architecture, with Atelier Kempe Thill - Architects and Planners, Rotterdam (NL), and Transver, Munich (D)

第二轮
4 AllesWirdGut Architektur, Vienna (A), with club L94 Landschaftsarchitekten, Cologne (D)
5 gmp International, Hamburg (D), with ST raum a. Landschaftsarchitekten, Berlin (D)
6 AS&P Albert Speer & Partner, Frankfurt am Main (D), with ver.de Landschaftsarchitektur, Freising (D)

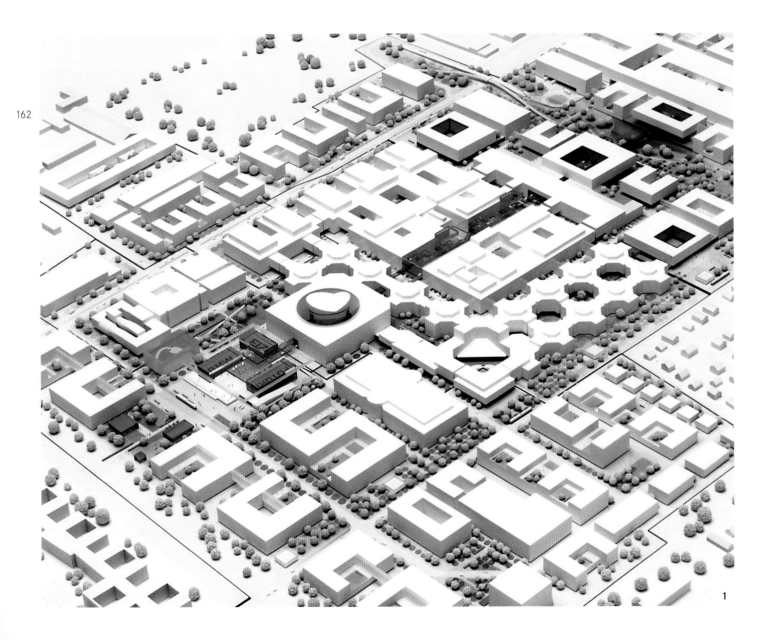

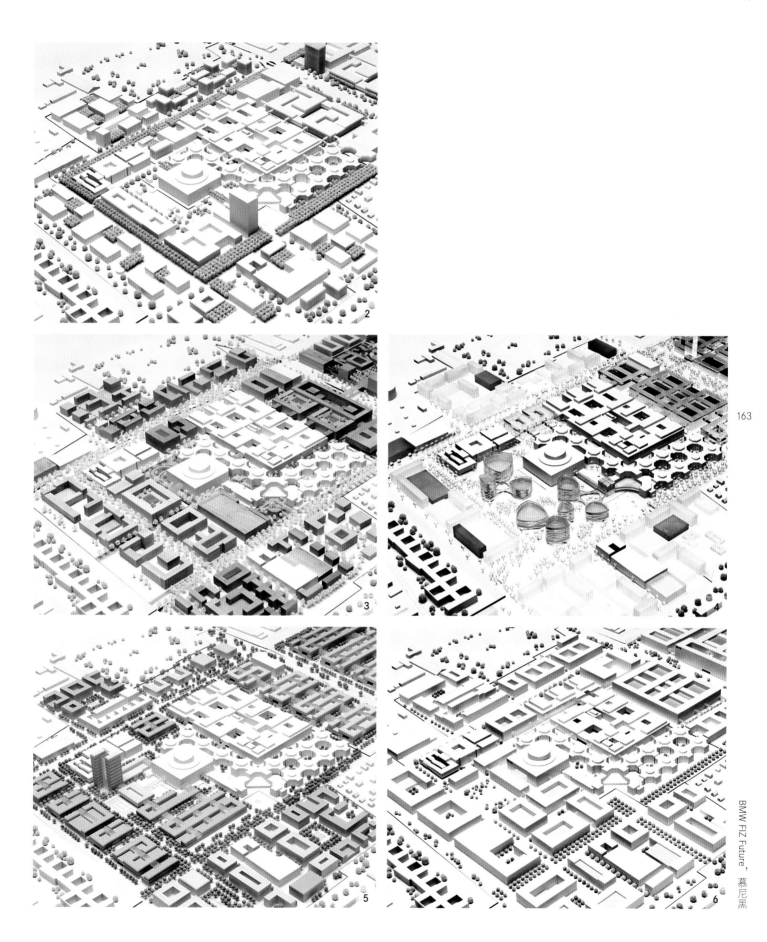

竞赛一等奖　获奖事务所

Henn Architekten

Munich (D); with Topotek 1, Berlin (D)

方案设计者：Gunter Henn, Martin Henn
参与者：Fredrik Werner, Patrick Br.II, Wolfgang Hirschmann, Maximilian Langwieder, Armin Nemati, Sascha Posanski
顾问：Regierungsbaumeister Schlegel, Munich (D); Lemon Consult, Zurich (CH); UBeG Dr. E. Mands & Dipl.-Geol. M. Sauer, Wetzlar (D)

Perspektive Magistrale

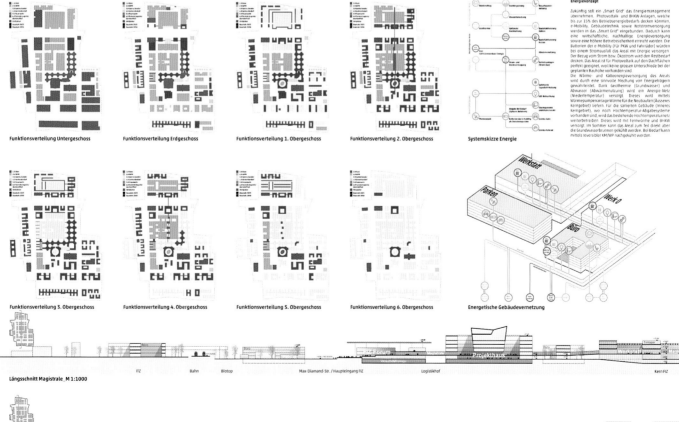

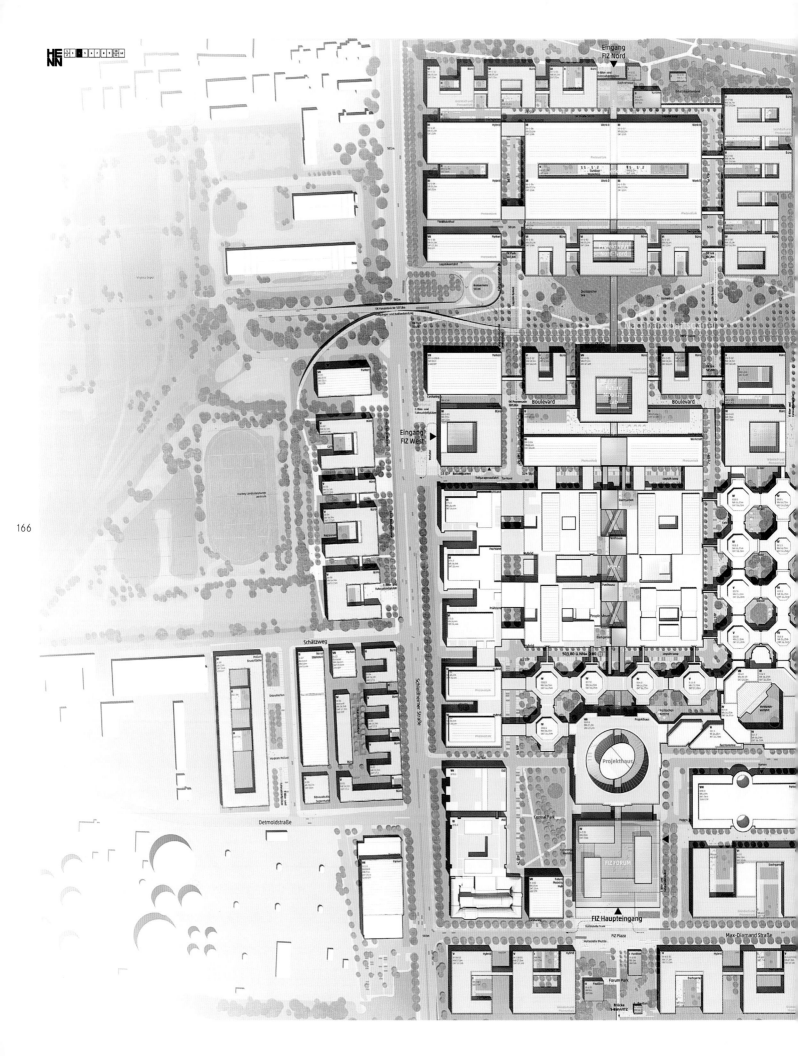

BEBAUUNGSSTRUKTUR

Hardware und Software
Das Fahrzeug steht im Mittelpunkt, die Querschnitts- und Entwicklungsfunktionen lagern sich um das Produkt an und überschneiden sich im Zentrum. Dieses bewährte Organisationsprinzip wird in der Erweiterung der Bebauung fortgeführt.

Mitte und Ränder
Im Zentrum konzentrieren sich jeweils die großmaßstäblichen Werkstätten und Hallenbereiche, hier entsteht auch die höchste bauliche Verdichtung. Die arbeitsplatzintensiveren Büro- und Gemeinschaftsflächen werden ringförmig um die zentralen Werkbereiche angeordnet. Die Randbebauung öffnet sich zu den umgebenden Landschaftsräumen, gleichzeitig schirmt sie die Umgebung vom internen Verkehr und Emissionen ab. Sie ist in ihrer Typologie und jeweiligen städtebaulichen Situation entsprechend deutlich gegliedert und vermittelt zur Nachbarschaft. Die „Harte Mitte" und die „Weichen Ränder" entsprechen im übertragenen Sinn der inhaltlichen Organisation des FIZ. Die Höhenentwicklung innerhalb des Masterplans reagiert ebenfalls differenziert auf die jeweilige städtebauliche Situation.

Kerngebiet und Cluster
Die Kernzone des FIZ Campus liegt zwischen Schleißheimer Straße, Knorrstraße, Max-Diamand-Straße und Rathenaustraße. Sie nimmt alle integrierenden Kernkompetenzen auf. In den umgebenden Bereichen können autarke Technologiecluster oder Kooperationspartner angesiedelt werden. Jedes Baugebiet wird durch eine klare Definition der Kanten bestimmt.

Adressen
Durch die Verlegung des Haupteinganges nach Süden in die Max Diamand Straße und die zwei zusätzlichen Zugänge im Westen an der Schleißheimer Straße und Osten an der Knorrstraße werden neue Adressen definiert. Sie bestimmen die Grundstruktur des Standortes. Jede neue Adresse erhält deutlich ablesbare Vorbereiche mit öffentlichen Funktionen und Mobilitätsangeboten.

FIZ Forum
In der Max Diamand Straße wird vor dem südlichen Haupteingang ein städtischer Platz ausgebildet (FIZ Plaza) der als urbaner Knotenpunkt zwischen Stadt, Nachbarschaft und Unternehmen fungiert und sich nach Süden als Forum Park zum Bahndammbiotop erweitert. Das FIZ Forum als multifunktionales Eingangsgebäude stellt den Auftakt für die interne Magistrale und die Schnittstelle zur Öffentlichkeit dar.

Zugänge West und Ost
Die Eingänge „West" und „Ost" an der Knorrstraße geben dem FIZ Campus ein Gesicht zur Schleißheimer und zur Knorrstraße. Es entstehen öffentliche Räume, die gemeinsam mit der Nachbarschaft genutzt werden und mit ihr in Dialog treten. Die Zugangsgebäude erlauben mit steuerbaren Eindringtiefen den Austausch und die Kooperation mit Externen. Der bestehende Haupteingang wird weiter als leistungsfähiger Zugang für Mitarbeiter von der Knorrstraße genutzt.

STADTRAUM

Raumbildung
Im Osten entlang der Knorrstraße und im Norden entlang der Rathenaustraße werden die Gebäude bewusst zurückgesetzt, um Abstand zu gewinnen und hochwertige öffentliche Außenräume zu gestalten. Ähnlich wird mit den Clustern östlich der Knorrstraße verfahren. Die Schleißheimer Straße wird durch Baumreihen vor einer beidseitigen Raumkante zur städtischen Allee. Die Max Diamand Straße wird durch den Vorplatz und Haupteingang zur urbanen Querverbindung. Der neu angelegte Nachbarschaftsgarten stellt die Verbindung zur übergeordneten Grünzone her und ermöglicht die Durchquerung der Kernzone. Durch die Ausnutzung des Terrainversprungs kann sowohl die öffentliche Querung sichergestellt werden, als auch die zwingende innere Einheit der Kernzone. Im neuen Park sind Cafe, Spiel- und Sportplätze, Bürgerzentrum und Verweilplätze für die Nachbarschaft vorgesehen.

Öffentliche Funktionen
Die Zugänge zum FIZ bieten Raum für öffentliche Nutzungen an. Ebenso werden in den umgebenden Clustern Quartiersplätze ausgebildet an denen sich öffentliche Funktionen anlagern können. Entlang der baumbestandenen Promenade an der Knorrstraße schaffen erdgeschossige Nutzungen und Pavillons ein zusätzliches Angebot. Die Aktivierung des Stadtraumes kommt Nachbarschaft und Unternehmen gleichermaßen zu Gute und schafft eine urbane Attraktivität.

Kernziele des Verkehrskonzeptes
- Störungsminimierte Steuerung und Entzerrung des Verkehrsaufkommens.
- Einbindung und Vernetzung des FIZ mit dem öffentlichen Nahverkehr.
- Integration innovativer Mobilitätskonzepte.
- Kurze Wege zum Arbeitsplatz.
- Entlastung der Wohngebiete, Verkehrsberuhigung der Knorrstraße.
- Gezielte Förderung der Fahrraderschließung, Anschluß an übergeordnete Fahrradwegesysteme.

Mobilitätsdrehscheiben
Die Zugangsbereiche werden als Mobilitätsdrehscheiben ausgebildet, an denen eine Überlagerung und Verdichtung verschiedener Verkehrsmittel erfolgt. Die Zugänge liegen an den Schwerpunkten der arbeitsplatzintensiven Zonen des FIZ und ermöglichen eine schnelle und direkte Verbindung zum Arbeitsplatz. Carsharing und E-Mobility konzentrieren sich an den Drehscheiben im Süden und Westen.

Parken
Durch die gezielte Anordnung der Parkhäuser und Garagen und direkte Zuordnung zu den Zugängen und den arbeitsplatzintensiven Bereichen im FIZ erfolgt eine Entzerrung des PKW Verkehrs und eine schnelle und unmittelbare Erschließung für Mitarbeiter und Besucher. Die Zufahrten der Parkhäuser erhalten entsprechende Rückstaubereiche.
Im Bereich der nördlichen Knorrstraße und am Zugang Ost gibt es kein Angebot an Parkmöglichkeiten, um die gewünschte Verkehrsberuhigung zu den angrenzenden Wohngebieten zu ermöglichen.

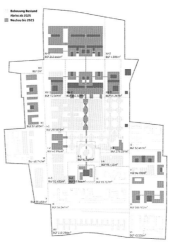

Ausbaustufe 2025

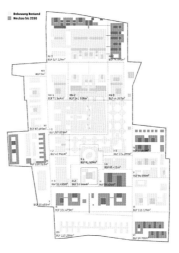

Ausbaustufe 2050

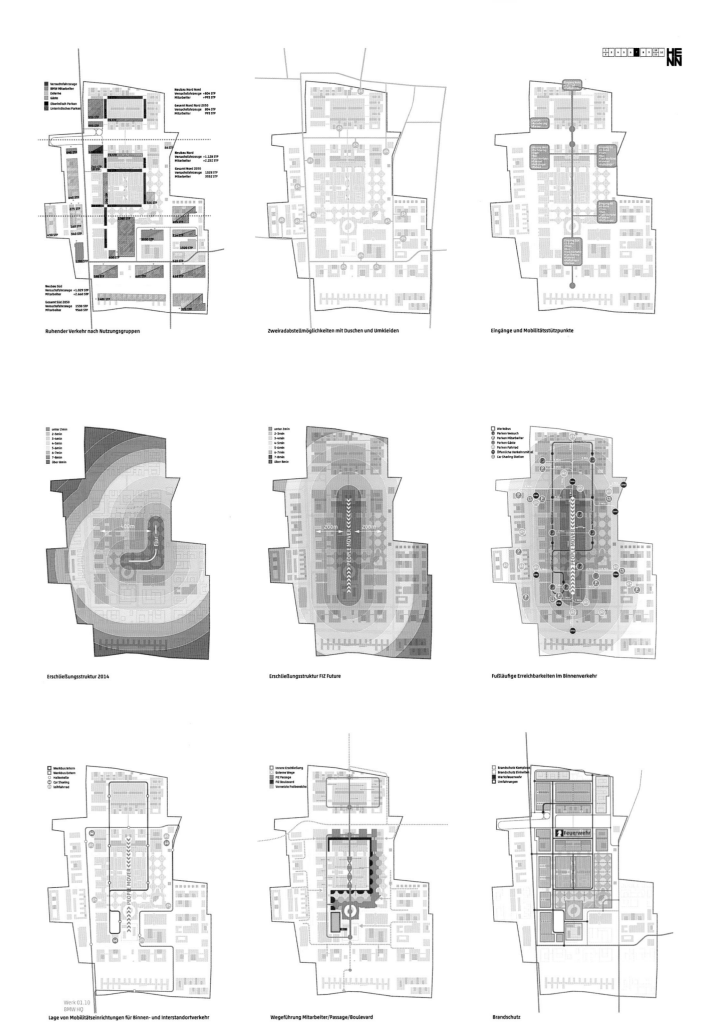

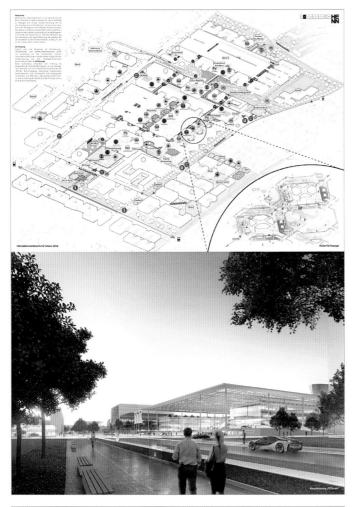
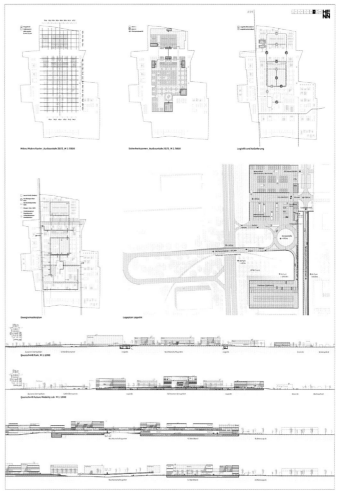
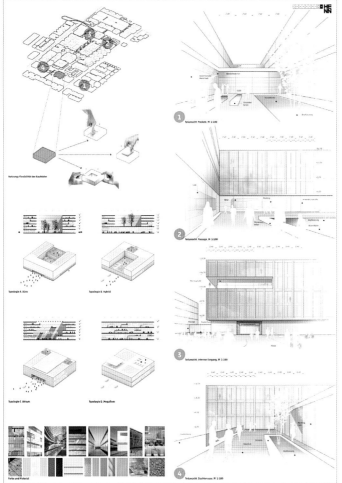
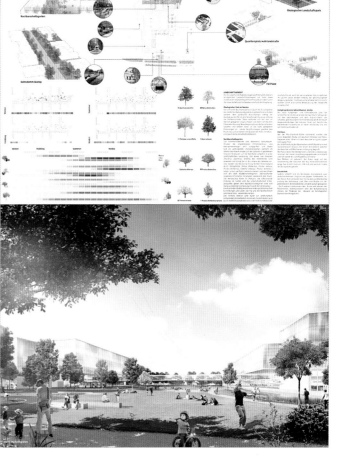

Henn Architekten — BMW FIZ Future, 慕尼黑

竞赛二等奖 获奖事务所

ernst niklaus fausch architekten

Zurich (CH); with Müller Illien Landschaftsarchitekten, Zurich (CH); Rapp Infra, Basel (CH)

方案设计者： Ursina Fausch
参与者： Bertram Ernst, Erich Niklaus, Lena Jung, Simone Cartier, Mireia Aixelà Bohigas, Bryan Graf, Klaus Müller, Rita Illien, Emmanuel Tsolakis, Beni Strub, Stefan Schneider
顾问： Planungsbüro Michael Angelsberger, Munich (D); Dr. Lürchiger & Meyer, Bauingenieure, Zurich (CH); Waldhauser + Hermann, Münchenstein (CH); Palis Projects & Engineering, Gersthofen (D); nightnurse images, Zurich (CH); Knecht Partner Modellbau, Wettingen (CH)

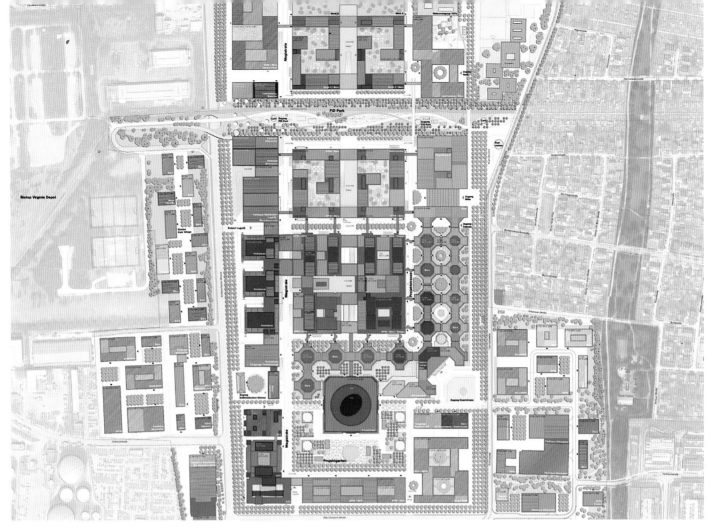

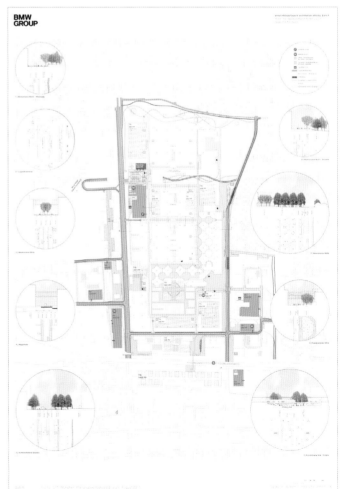

BMW GROUP

ernst niklaus fausch architekten eth/sia, Zürich
Müller Illien Landschaftsarchitekten, Zürich
Rapp Infra AG, Basel

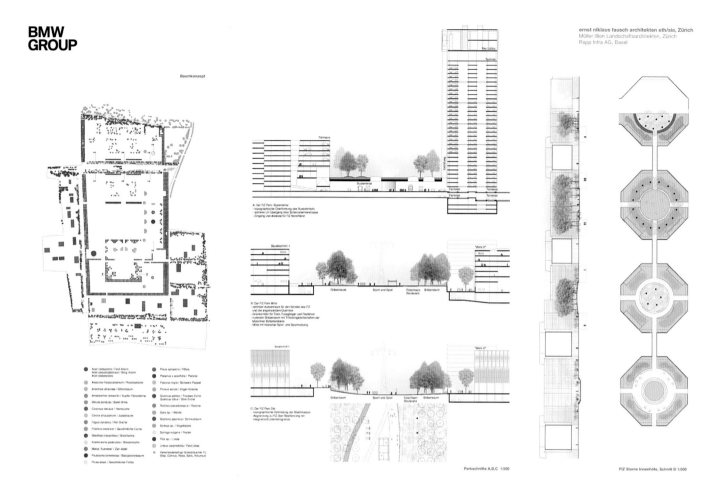

Der Projektgarten repräsentiert das Leben und Arbeiten auf dem FIZ-Campus für Mitarbeiter und deren Besucher. Auf Augenhöhe mit Kunst und Kultur bietet er vielschichtige Orte der Kommunikation.

Blatt 10 + 11 FIZ FUTURE München, Planungswettbewerb 2. Stufe, August 2014 — Freiraumgestaltung

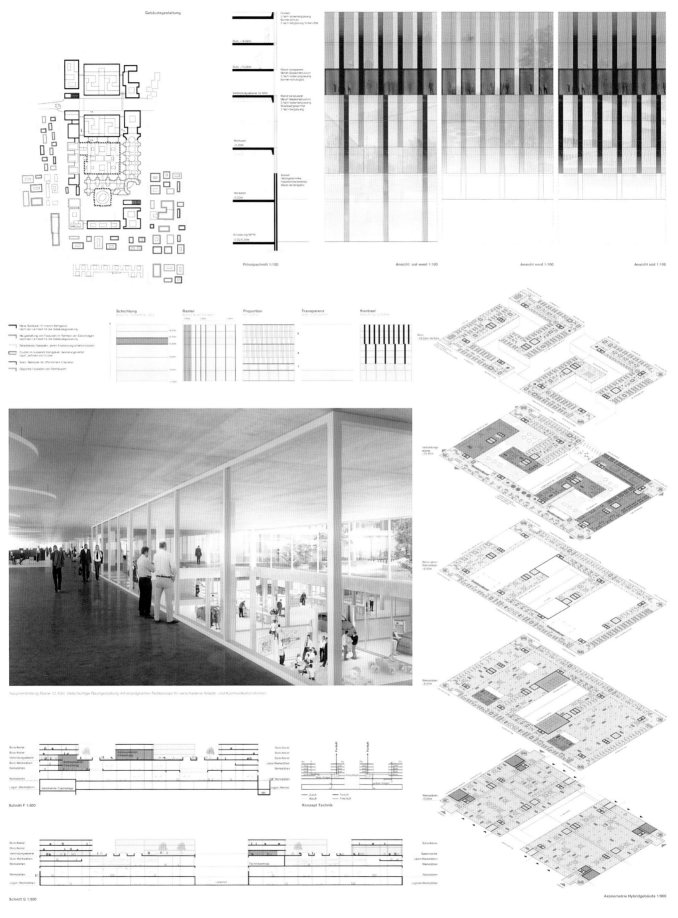

竞赛三等奖 获奖事务所

West 8 Urban Design & Landscape Architecture

Rotterdam (NL); with Atelier Kempe Thill - Architects and Planners, Rotterdam (NL); Transver, Munich (D)

方案设计者：Adriaan Geuze, André Kempe, Jürgen Schmiele
参与者：Christoph Els.sser, Simone Huijbregts, Janneke Eggink, Karsten Buchholz, Bruno W.ber, Ben Wegdam, Giulia Frittoli, Saskia Hermanek, Anne-Laure Gerlier, Pauline Burand, Thomas Antenner, Laura Paschke
自由职业者：Daniela Bergmann

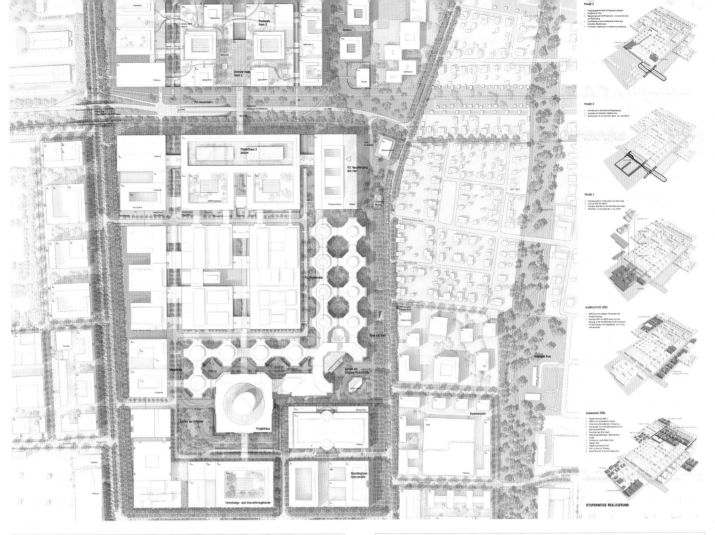

STUFENWEISE REALISIERUNG

FIZ FUTURE ARBEITSWELTEN

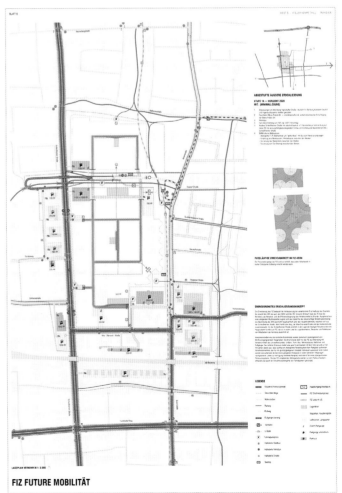

FIZ FUTURE MOBILITÄT

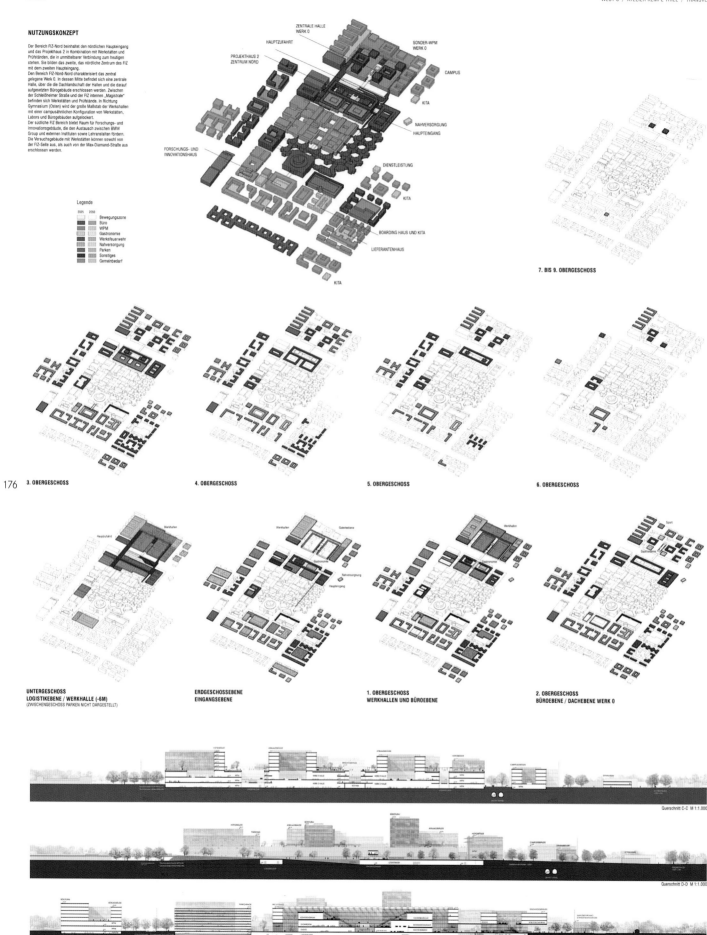

BLATT 10

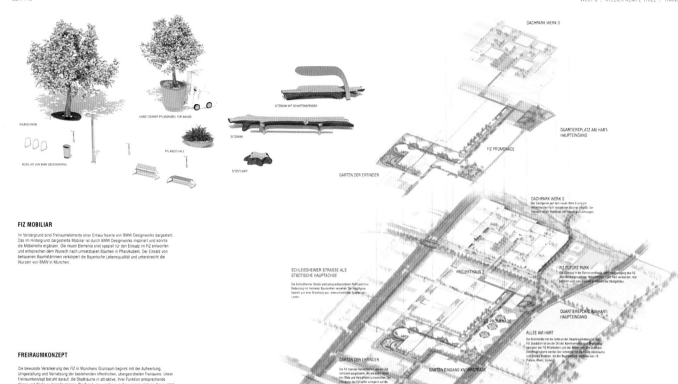

FREIRAUMKONZEPT

FIZ MOBILIAR

Im Vordergrund sind Freiraumelemente einer Entwurfsserie von BMW Designworks dargestellt. Das im Hintergrund dargestellte Mobiliar ist durch BMW Designworks inspiriert und könnte die Möbelreihe ergänzen. Die neuen Elemente sind speziell für den Einsatz im FIZ entworfen und entsprechen dem Wunsch nach umsetzbaren Bäumen in Pflanzkübeln. Der Einsatz von behauenen Baumstämmen verkörpert die Bayerische Lebensqualität und unterstreicht die Wurzeln von BMW in München.

FREIRAUMKONZEPT

Die bewusste Verankerung des FIZ in Münchens Grünraum beginnt mit der Aufwertung, Umgestaltung und Vernetzung der bestehenden öffentlichen, übergeordneten Freiräume. Unser Freiraumkonzept beruht darauf, die Stadträume in attraktive, ihrer Funktion entsprechende Alleen und Parks zu transformieren. Die Freiräume werden aus ihrer gegenwärtigen Anonymität befreit und werden zu Treffpunkten und Orten der Kommunikation. Sie werden entsprechend ihrer funktionalen Bedeutung unterschiedlich gestaltet und bieten hochwertige Lagen. Das Freiraumkonzept des Kerngebietes orientiert sich ähnlich der Funktionsverteilung an den werksinternen Mitarbeiterströmen. Die Freiräume haben die Aufgabe Mitarbeiter zusammenzuführen und ihren Gedankenaustausch zu fördern.

LEITBAUMKONZEPT

Bäume sind Identität stiftende Elemente im Stadtgefüge und tragen zur Adressbildung bei. Die Baumwahl ist von der Bedeutung des jeweiligen Straßen- oder Grünraumes in der Stadtstruktur abhängig. Das FIZ und seine Umgebung sollen von einer Vielzahl unterschiedlicher Baumsorten geprägt sein, die den Stadträumen jeweils ihren unverwechselbaren Charakter geben.

BLATT 11

DIE GÄRTEN DER ERFINDER
Erholen, entspannen und verweilen beim südlichen Projekthaus

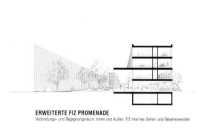

ERWEITERTE FIZ PROMENADE
Verbindungs- und Begegnungsraum, innen und Außen; FIZ internes Sehen und Gesehenwerden

INTERNER BEGEGNUNGSRAUM
Zentrales Atrium Projekthaus 2, gebäudeinterner Außenraum.

STADTALLEE KNORRSTRASSE

FIZ PROMENADE
Aufwertung der östlichen Achse entlang der Sterne zum internen und externen Begegnungsraum

BEGEGNUNGSRÄUME

第二轮 获奖事务所

AllesWirdGut

Vienna (A); mit club L94 Landschaftsarchitekten, Cologne (D)

方案设计者： Christian Waldner, Burkhard Wegener
参与者： Antonio Boeri,
Felix Reiner, Michael Stehlik, Ondrej Stehlik, Andreas Junges, Vera Pistkova, Franziska Lesser
顾问： PSLV Planungsgesellschaft Stadt-Land-Verkehr, Munich (D); Transsolar Energietechnik, Stuttgart (D)

FIZ FUTURE MÜNCHEN

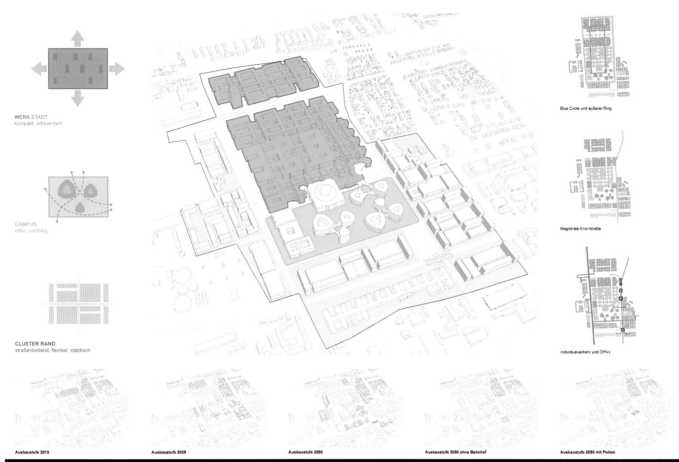

AllesWirdGut & Club L94

第二轮 获奖事务所

gmp International

Hamburg (D); mit ST raum a. Landschaftsarchitekten, Berlin (D)

方案设计者： Nikolaus Goetze, Stefan J.ckel
参与者： Marc Ziemons, Jan Blasko, Friedrich Prigge, Tim Leimbrock, Yping Tang, Christoph Berle, Tom Schuelke, Markus Carlsen, Urs Wedekind, Jia Zhang, Mathias Werner
顾问： ARGUS, Stadt- und Verkehrsplanung, Hamburg (D); RMN Ingenieure, Hamburg (D)

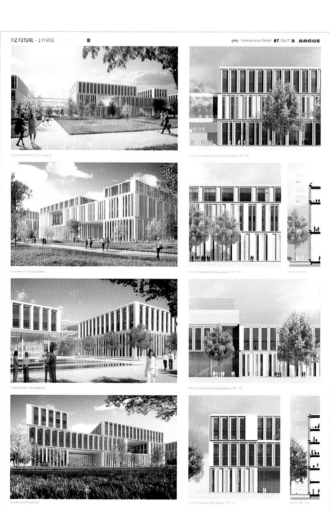
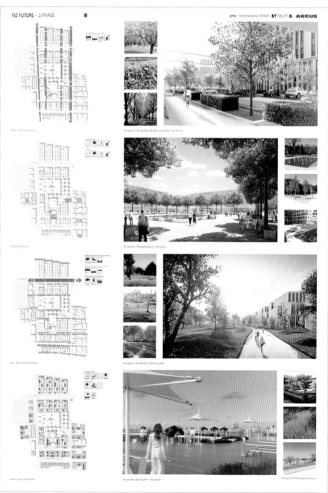

BMW FIZ Future · 慕尼黑

第二轮　获奖事务所

AS&P Albert Speer & Partner

Frankfurt am Main (D); mit ver.de Landschaftsarchitektur, Freising (D)

方案设计者：Albert Speer, Birgit Kr.niger
参与者：Michael Heller, Michael Dinter, Daniel Ringeisen, Pascal Kuhn, Andrea Mandic, Frederik Lux, Caroline Morell, Stephan Gentz, Maike Eggeling
顾问：Transsolar Energietechnik, Stuttgart (D)

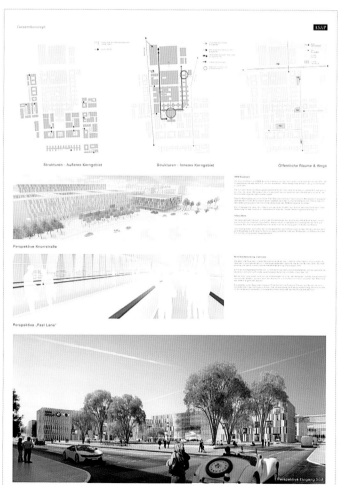
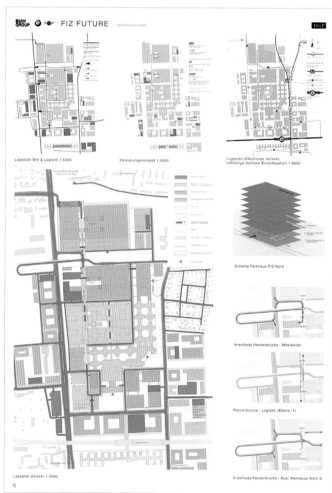

威斯巴登法学院
（隶属于欧洲商业学校）

德国 威斯巴登

2009—2010

概况

业主：海塞省物业管理和施工管理局西部分局
项目规模：基地面积 11 300 平方米
类型：双阶段开放型设计竞赛（开放申请程序）
参赛者：第一阶段 22 组，第二阶段 7 组
竞赛预算：222 000 欧元（奖金 138 000 欧元，第二阶段每组参赛者的费用 15 000 欧元）

评委

建筑评委：Prof. Mario Campi, Architect, Lugano/Zurich; Prof. Rebecca Chestnutt, Architect, Berlin/Stuttgart; Harald Clausen, Undersecretary and Architect, Hessische Baumanagement, Wiesbaden; Markus Karow, Architect, Offenbach; Dirk Neujahr, Architect, European Business School, Oestrich-Winkel; Volker Staab, Architect, Berlin

专家评委：Elmar Damm, Head of Section, Ministry of Finance of Hessen, Wiesbaden; Georg Engel, Building Director, Hessisches Immobilienmanagement, Wiesbaden; Prof. Dr. Christopher Jahns, President of the European Business School, Oestrich-Winkel; Dr. Reimar Palte, Chancellor of the European Business School, Oestrich-Winkel; Prof. Dr. Joachim Pös, Department Head Urban Development and Transport, Wiesbaden

"竞赛任务是建造一栋示范性大学建筑，结合场地内的历史建筑，并从形态、功能、结构及操作的经济性等方面清晰地表现建筑的当代性和可持续性。"

——引自《竞赛摘要》

威斯巴登法学院

威斯巴登是德国海塞省首府及议会所在地。威斯巴登的历史中心和温泉度假地几乎完全免受战争的破坏，因此，威斯巴登被认为是一座极具历史意义的19世纪城市。海塞省和威斯巴登市政府的总体规划是让威斯巴登成为一座大学城。2009年的规划要求在欧洲商业学校中建造一座法学院。欧洲商业学校是德国历史最悠久的私立工商管理大学，拥有高质量的师资队伍。得益于威斯巴登法学院的建成，欧洲商业学校成功地跻身欧洲法律领域领军大学行列。

场地以及旧城区的法院大楼

建于1897年的区法院和建于1867年的官邸定义了城市中心的场地特征，它们共同形成了特殊而单一的空间状态，是始于1851年的沿莱恩街城市拓展规划的一部分。空置建筑的再利用成为该项目的重要组成部分。

竞赛任务

建成后的威斯巴登法学院包括11 500平方米的演讲厅、教学区、图书馆和管理办公室。此外，不同于公立大学，多功能会议室可举办培训及相关活动，这是学校经费的重要来源。现有法院大楼可用空间占项目用地约50%。建筑师特别关注三个主题：尊重现有建筑的规划布局，在用地范围内随地形变化灵活组织，在中心位置形成一个特色区域。此外，在法院用地范围内专门开辟了一方土地，专供历史纪念之用，因其在1936—1945年间曾作为盖世太保监狱。

威斯巴登法学院（隶属于欧洲商业学校），威斯巴登

第一阶段参赛者

第二阶段入围者

1　gmp Generalplanungsgesellschaft, Frankfurt am Main (D)
2　Eller + Eller Architekten, Düsseldorf (D)
3　msm meyer schmitz-morkramer, Darmstadt (D)
4　Nieto Sobejano Arquitectos, Berlin (D)
5　bgf+ architekten, Wiesbaden (D)
6　3XN A/S, Copenhagen (D)
7　Zinterl Architekten, Graz (A)

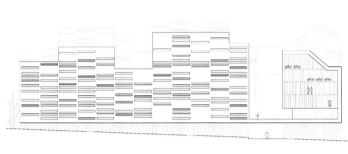
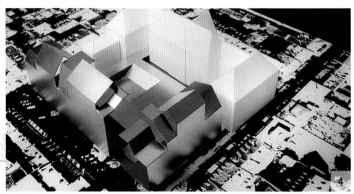

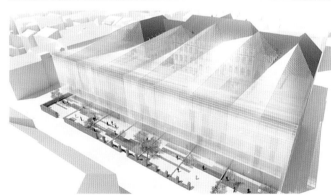

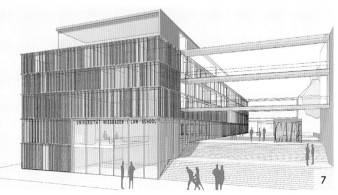

第一阶段其他参与者

8 Max Dudler Architekten, Berlin (D)
9 Gerber Architekten, Dortmund (D)
10 MGF Architekten, Stuttgart (D)
11 Baumschlager Eberle Gruppe | Baumschlager Eberle Lochau, Lochau (A)
12 Böge Lindner Architekten, Hamburg (D)
13 Gatermann + Schossig Bauplanungsgesellschaft, Cologne (D)
14 Turkali Architekten, Frankfurt am Main (D)
15 Kuehn Malvezzi, Berlin (D)
16 Auer + Weber + Assoziierte, Stuttgart (D)
17 kadawittfeldarchitektur, Aachen (D)
18 Schweger Associated Architects, Hamburg (D)
19 Rapp+Rapp, Berlin (D)
20 Ferdinand Heide Architekt, Frankfurt am Main (D)
21 Weber Hofer Partner Architekten, Zurich (CH)
22 BIG - Bjarke Ingels Group, Copenhagen (D)

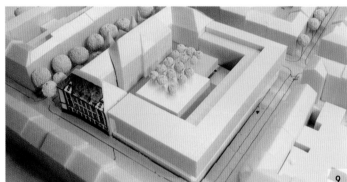

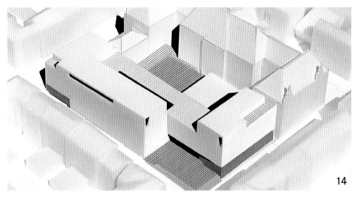

14

15

16

17

18

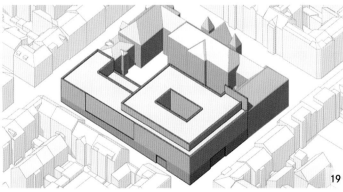

19

20

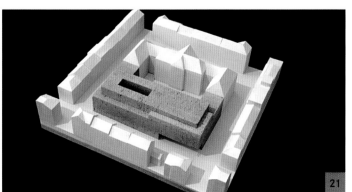

21

22

第二阶段参赛者

获奖者

1. 一等奖 Eller + Eller Architekten, Düsseldorf (D)
2. 二等奖 Zinterl Architekten, Graz (A)
3. 三等奖 3XN A/S, Copenhagen (D)
4. 四等奖 Nieto Sobejano Arquitectos, Berlin (D)

第二轮

5. msm meyer schmitz-morkramer, Darmstadt (D)
6. gmp Generalplanungsgesellschaft, Frankfurt am Main (D)
7. bgf+ architekten, Wiesbaden (D)

竞赛一等奖　获奖事务所

Eller + Eller Architekten

Düsseldorf (D)

方案设计者： Erasmus Eller
参与者： Martin Schliefer, Joern Lammert
自由职业者： Christiane Flasche, Achille Farese
顾问： Scholze Ingenieurgesellschaft, Leinfelden-Echterdingen (D); St raum a Landschaftsarchitekten, Berlin (D), Tobias Micke; Krebs und Kiefer, Darmstadt (D)

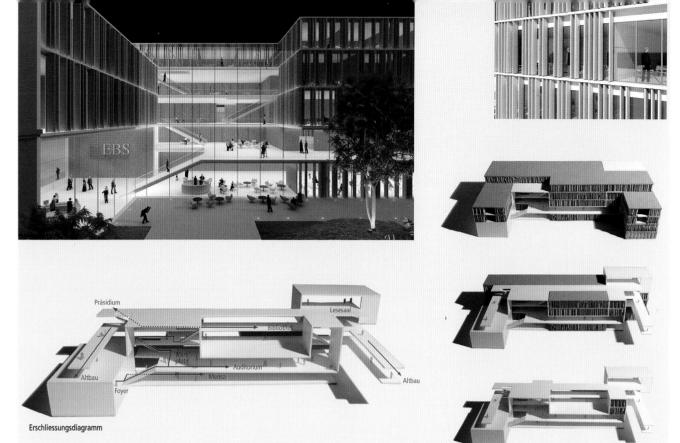

神经退行性疾病中心

德国 波恩

2011

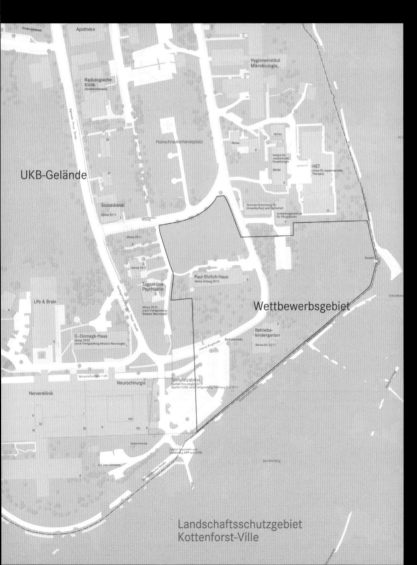

概况
业主：神经退行性疾病中心
项目规模：基地面积 15 400 平方米
类型：开放型设计竞赛（开放申请程序）
参赛者：12 组
竞赛预算：580 000 欧元（奖金 280 000 欧元，每组参赛者的费用 25 000 欧元）

评委
建 筑 评 委：Prof. Mario Campi, Architect, Lugano/Zurich; Günther Hoffmann, Architect, Director of Construction Department, Federal Ministry of Transport and Digital Infrastructure, Berlin; Olaf Rabe, Architect, Braunschweig; Prof. Kirsten Schemel, Architect, Berlin/Münster; Heiko Schiller, Building Services and Environmental Engineer, Hamburg; Ralf Streckwall, Architect Max Delbrück Center for Molecular Medicine (MDC), Berlin; Friedhelm Terfrüchte, Landscape Architect, Essen

专家评委：Bärbel Brumme-Bothe, Director, Federal Ministry of Education and Research (BMBF), Berlin; Prof. Dr. Michael Lentze, Universitätsklinikum Bonn; Prof. Dr. Pierluigi Nicotera M.D. Ph.D., German Center for Neurodegenerative Diseases (DZNE), Bonn; Dr. Beate Müller, Ministerium für Innovation, Wissenschaft und Forschung des Landes Nordrhein-Westfalen (MIWF-NRW), Düsseldorf; Jürgen Nimptsch, Lord Mayor of Bonn; Ursula Weyrich, German Center for Neurodegenerative Diseases (DZNE), Bonn

"竞赛任务是为德国神经退行性疾病中心建造一座新建筑，为生命科学领域中的科学研究和技术创新提供最佳空间，并促进常驻科学家与访问科学家的交流与合作。新建筑象征着德国神经退行性疾病中心的科研与技术成果已达到国际一流水平和先进等级，其外部和内部设计也应有所体现，至少不应有所逊色。"

——引自《竞赛摘要》

神经退行性疾病中心
即使在今天，神经退行性疾病的成因和药理作用在很大程度上还是未知的。神经退行性疾病中心致力于神经退行性疾病的基础研究和临床实践，推动该领域的科研发展与技术进步，提升德国在该领域的国际知名度。

项目所在地波恩
自 2009 年起，神经退行性疾病中心的总部和基础设施设在波恩。合作机构分布在德累斯顿、慕尼黑、哥廷根、马格德堡等城市。新建筑位于波恩医科大学维纳斯校区南部边缘。到 2020 年，新建筑的结构和功能将在总体规划的框架下逐年更新。总体规划的内容包括：重建校园南部区域（神经学研究中心、神经病学研究中心和疾病治疗中心），建立南部急救中心，修复和更新神经病学门诊部，与神经退行性疾病中心旧楼相串联。

竞赛任务
竞赛用地是神经退行性疾病中心总体规划的区域。占地 16 000 平方米的新建筑致力于为国际研究小组的科研工作（包括生物医学实验、临床研究领域、空间磁共振成像和人口研究）提供理想的条件。相对集中的研究设施和总部的管理部门被纳入新建筑。随着用地范围扩大，建筑师逐渐偏重于神经退行性疾病中心场地长期规划与长远潜力的比试与较量。

参赛者

获奖者

1 一等奖 wulf & ass. Architekten, Stuttgart (D)
2 二等奖 hammeskrause architekten, Stuttgart (D)
3 三等奖 SOW Planungsgruppe, Berlin (D)

第二轮

4 doranth post architekten, Munich (D)
5 gmp Generalplanungsgesellschaft, Aachen (D)
6 Nickl & Partner Architekten, Munich (D)
7 Henn Architekten, Munich (D)
8 Fritsch + Tschaidse Architekten, Munich (D)
9 Behnisch Architekten, Stuttgart (D)

第一轮

10 Henning Larsen Architects, Copenhagen (DK)
11 Rudy Uytenhaak Architectenbureau, Amsterdam (NL)
12 Heinle, Wischer Gesellschaft für Generalplanung, Cologne (D)

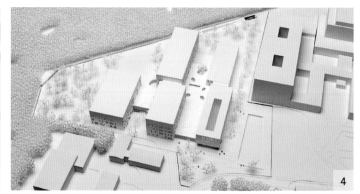
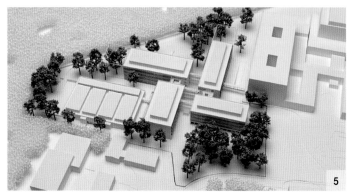
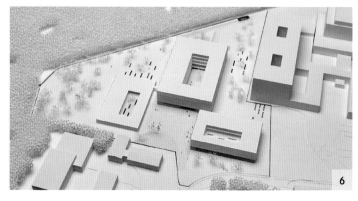

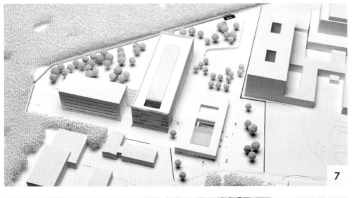
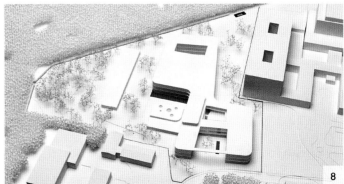

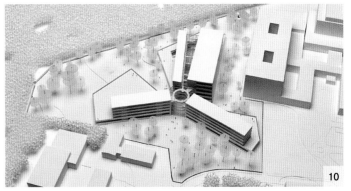
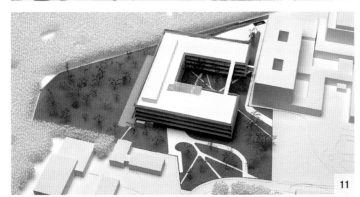

竞赛一等奖　获奖事务所

wulf & ass. Architekten

Stuttgart (D); with IWP Ingenieurbüro für Systemplanung, Stuttgart (D); Mayr|Ludescher|Partner Beratende Ingenieure, Stuttgart (D); Adler & Olesch Landschaftsarchitekten, Stadtplaner SRL und Ingenieure, Mainz (D)

方案设计者：Tobias Wulf
参与者：Steffen Vogt, Andreas Moll, Camilo Hernandez, Boris Weix, Violett Kratzke, Gaston Stoff, Meike Zwerger

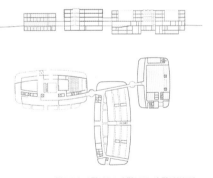

DEUTSCHES ZENTRUM FÜR NEURODEGENERATIVE ERKRANKUNGEN _ BONN

竞赛二等奖　获奖事务所

hammeskrause architekten

Stuttgart (D); with IGF Ingenieurgesellschaft Feldmeier, Münster (D);
GLP Ingenieurgesellschaft, Hamburg (D);
Dr. Heinekamp Labor- und Institutsplanung, Berlin (D);
Gantert + Wiemeler Ingenieurplanung, Münster (D);
Eurich Gula Landschaftsarchitekten, Wendlingen (D)

方案设计者：Markus Hammes, Nils Krause
参与者：Uwe Beierbach, Claudia Büchler, Peter Just, Astrid Karr, Ajla Pasic, Lima Feng

Zonierung

Eingang

Orientierung

Horizontale Vernetzung

Vertikale Vernetzung

Konzeptskizzen

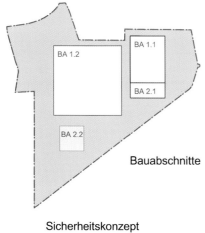

Bauabschnitte

Sicherheitskonzept

Nutzungs- und Zirkulationskonzept

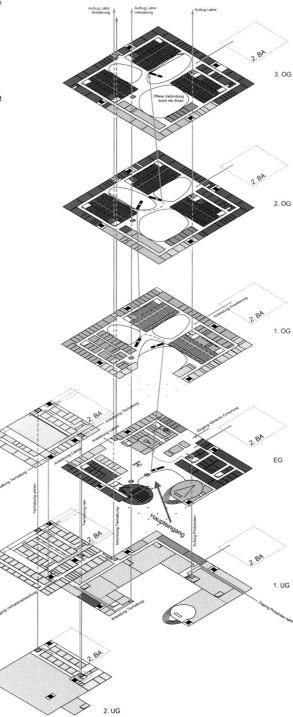

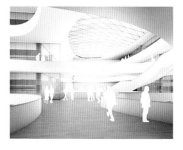

Funktionale Bezüge
Wissenschaftliche Bereiche

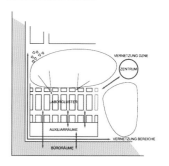

Funktionale Bezüge
Casino / Cafeteria / Küche

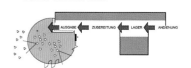

竞赛三等奖　获奖事务所

SOW Planungsgruppe

Berlin (D); with WBP Ingenieure für Haustechnik, Münster (D); osd, Frankfurt am Main (D); Levin Monsigny Landschaftsarchitekten, Berlin (D); IBB Ingenieurbüro, Leipzig (D)

方案设计者：Volker Staab
参与者：Petra W.Idle, Henriette Siegert, Diana Saric, Ivan Kaleov, Carolin Kuhn, Georg Hana, Noah Grunwald, Claudia Trott
顾问：Prof. Gert Beilicke, Marco Schm.ller
其他顾问：LCI Labor Concept Ingenieurgesellschaft, Lüneburg (D), Marco Pleu., Michael Goszdziewski

Deutsches Zentrum für Neurodegenerative Erkrankungen, Bonn

尤里西生物研究所
生物研究园

德国 尤里西

2014

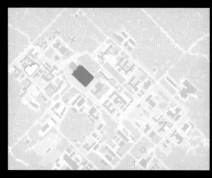

概况

业主：尤里西生物研究所

项目规模：基地面积 3000 平方米

类型：开放型设计竞赛（开放申请程序）

参赛者：13 组

竞赛预算：130 000 欧元（奖金 48 000 欧元，荣誉奖 10 000 欧元，每组参赛者的费用 6000 欧元）

评委

建筑评委：Prof. Manfred Hegger, Architect, Kassel; Prof. Dörte Gatermann, Architect, Cologne; Alfred Nieuwenhuizen, Architect, Berlin; Olaf Rabe, Architect, Helmholtz Centre for Infection Research (HZI), Braunschweig; Ralf Streckwall, Architect, Head of Technical Facility Management, Max Delbrück Center for Molecular Medicine (MDC), Berlin; Daniel Wentzlaff, Architect, Basel

专 家 评 委：Karsten Beneke, Deputy Chief Executive Officer, Forschungszentrum Jülich GmbH; Prof. Dr. Michael Bott, Head of the Institute for Bio- and Geosciences (IBG-1), Forschungszentrum Jülich GmbH; Jens Kuchenbecker, Head of Planning and Building, Forschungszentrum Jülich GmbH; Prof. Dr. Wolfgang Marquardt, Chief Executive Officer Forschungszentrum Jülich GmbH; Gisela Nobis-Fritzen, Undersecretary, Federal Ministry for the Environment, Nature Conservation, Building and Nuclear Safety (BMBU)

"新建筑应该以创新且经济适用的形式为教学、研究和交流提供空间，并为可持续发展的终极目标提供明确的当代解答。"

——引自《竞赛摘要》

尤里西生物研究所

尤里西生物研究所是位于德国科隆和亚琛之间的研究中心，致力于亟待解决的跨学科生物研究，并附带一些可持续能源和环境以及信息和人体大脑方面的研究。尤里西生物研究所拥有超过 5000 名员工，它是欧洲主要跨学科研究中心——亥姆霍兹协会的成员之一。各分支机构及基础设施的整合是 2009 年尤里西生物研究所重组计划的一部分。

生物研究园

项目总体规划囊括了尤里西生物研究所整个场地未来发展的结构框架和长远战略，涉及若干个研究部门，每个部门均位于生物研究园的一座建筑中。生物研究园位于尤里西生物研究所西北边缘，基地上屹立着几座建筑。此次规划对生物研究园的空间布局做了完整的梳理，将现有建筑进行重组的同时较和谐地融入了区域内的自然景观。

竞赛任务

竞赛任务是建造一座具有开创性的实验办公楼，面积约 4400 平方米，来自三个部门的研究人员在此办公。尤里西生物研究所希望其所具有的极高的学术水平在实验办公楼内外部设计和材料选取中有所体现。除单一部门的实验和科研活动外，实验办公楼也是多学科交叉且组织复杂的众多科研机构的日常科研之地，致力于各科研机构在不同研究层面上和不同学科之间的技术交流与成果共享。

尤里西生物研究所生物研究园，尤里西

参赛者

获奖者

1 一等奖 Rohdecan Architekten, Dresden (D)
2 二等奖 Atelier 30 Architekten, Kassel (D)
3 三等奖 Hascher Jehle Generalplanungsgesellschaft, Berlin (D)
4 荣誉奖 wulf architekten, Stuttgart (D)
5 荣誉奖 Glass Kramer L.bbert Architekten with Itten+Brechbühl and P.arc, Berlin (D)

第二轮

6 gmp International, Aachen (D)
7 Brullet - De Luna Arquitectes with Pinearq, Barcelona (E)
8 Behnisch Architekten, Stuttgart (D)
9 BHBVT Architekten, Berlin (D)
10 Riegler Riewe Architekten, Graz (A)
11 Baumschlager Eberle Gruppe | BE Berlin, Berlin (D)
12 W.rner Traxler Richter Planungsgesellschaft, Frankfurt am Main (D)
13 Pich-Aguilera Arquitectos, Barcelona (E)

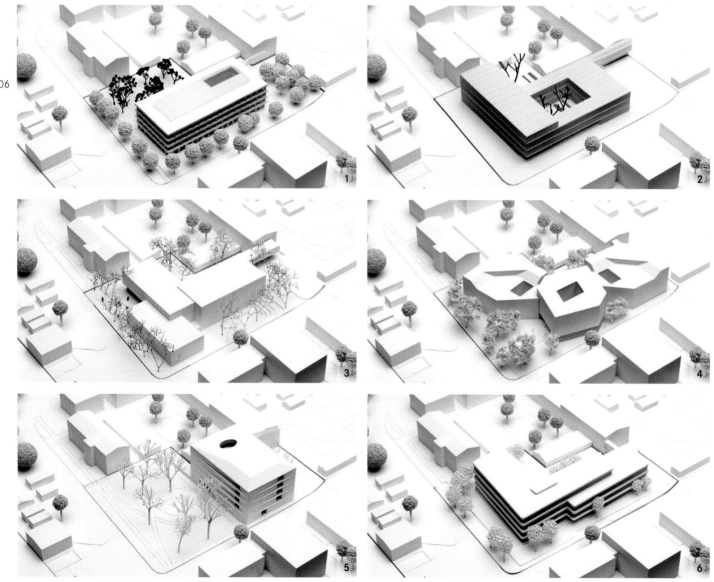

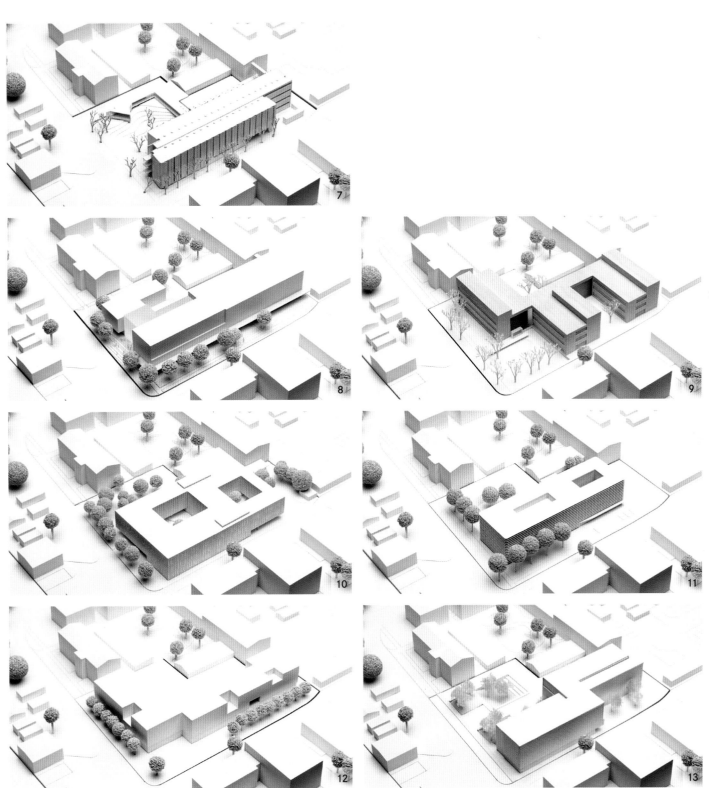

竞赛一等奖　获奖事务所

Rohdecan Architekten

Dresden (D); with INNIUS GTD, Dresden (D); IPN Laborprojekt, Dresden (D); Rehwaldt Landschaftsarchitekten, Dresden (D)

方案设计者：Eckart Rohde, Canan Rohde-Can
参与者：Sebastian Schr.ter, Caroline S.misch, Johannes Bürger

Nutzungs- und Zirkulationskonzept

Fassadendetail 1:25

Ansicht Süd-Ost (Labore) 1:200

Ansicht Süd-West (Labore) 1:200

竞赛二等奖　获奖事务所

Atelier 30 Architekten

Kassel (D); with Gnüchtel Triebswetter Landschaftsarchitekten, Kassel (D); ZWP Ingenieur, Wiesbaden (D)

方案设计者： Ole Creutzig, Thomas Fischer
参与者： Eva-Sophia Bisdorf, Yunus Coskun, Maria Eckstein, Katharina Port, Anika Schmidt, Georgios Varelis

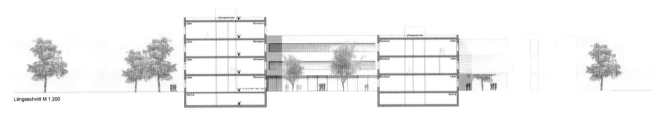

Längsschnitt M 1:200

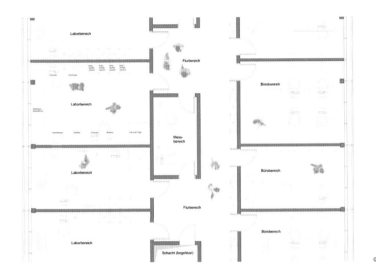

Grundriss Musterlabor M 1:50

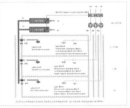

Systemschnitt Lüftung

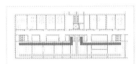

Systemgrundriss Lüftung

Für belastete Abluft gibt es u.U. keine bauaufsichtlich zugelassene Brandschutzklappen.
Zur Vermeidung von Brandschutzklappen in Abluftkanälen werden die Abluftstränge je Brandabschnitt in einer eigenen F90 abgetrennten „Kammer" innerhalb des Schachtes geführt.
Die Zuluft hingegen wird über einen Zentralschacht mit etagenweisen Abgängen mit Brandschutzklappe verteilt.

Eingangshof

竞赛三等奖　获奖事务所

Hascher Jehle Generalplanungsgesellschaft

Berlin (D); with Ingenieurbüro Mayer, Ottobeuren (D);
Teamplan, Tübingen (D)

方案设计者： Prof. Rainer Hascher, Prof. Sebastian Jehle
参与者： Stephanie Larassati, Agnese Di Quirico, Johannes Raible

Haupteingang

Drei ineinander verschachtelte kubische Baukörper unterschiedlicher Höhe gruppieren sich um das kommunikative Zentrum des Gebäudes. Der Gebäudeversatz am West-Ring formt einen kleinen Vorplatz und bildet den Haupteingang auf selbstverständliche Weise aus. Der Rücksprung der Erdgeschosszone verstärkt die einladende Geste.

Kommunikatives Zentrum

Im Zentrum des Gebäudes befindet sich ein offener Bereich für die Kommunikation der Mitarbeiter. Der Bereich dient als **TREFFPUNKT UND VERTEILER** des Gebäudes. Über die Gebäudefuge wird der **ATTRAKTIVE AUFENTHALTSBEREICH** belichtet.

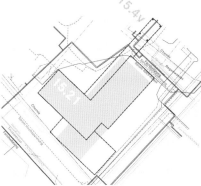

Bestandsleitungen

Der Neubau wird so auf dem Wettbewerbsgebiet positioniert, dass **ALLE BESTEHENDEN LEITUNGEN** unverändert **ERHALTEN** bleiben. Lediglich im Anschlussbereich des Bestandsgebäudes wird eine punktuelle Gründung für die Brückenkonstruktion, die das Gebäude 15.4.v anbindet, erforderlich. Kostenaufwendige Verlegungen der Bestandsleitungen werden vermieden.

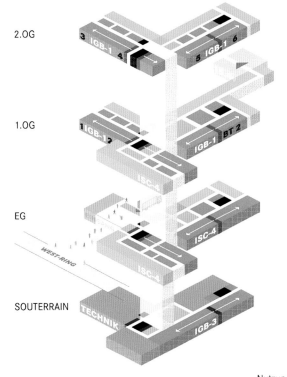

Die Gebäudeorganisation ähnelt der **STRUKTUR EINES BAUMES:**
Über die zentral gelegene Vertikalerschließung sind alle Institute direkt an die Eingangshalle angebunden. An der Vertikalerschließung im Zentrum des Gebäudes befinden sich in jedem Geschoss jeweils die gemeinsam genutzten Räume sowie die Kommunikationsbereiche, die Aufenthalts- und Pausenflächen für die Mitarbeiter. Um dieses kommunikative Zentrum herum gruppieren sich die jeweiligen Unterabteilungen geschossweise wie Blätter an einem Baum.
Die drei Institute werden horizontal über die Ebenen des Neubaus verteilt. Dadurch können die jeweiligen **INSTITUTE** größtenteils **EBENENGLEICH ORGANISIERT** werden.

Der Haupteingang des Neubaus befindet sich auf der Westseite des Wettbewerbsgebietes. Das Gebäude wird vom West-Ring aus erschlossen. Mitarbeiter und Besucher werden von einem **GROSSZÜGIGEN FOYER** empfangen. Die Eingangshalle lässt sich bei Bedarf durch den benachbarten Seminarraum zu einer größeren zusammenhängenden Fläche erweitern, die für **GRÖSSERE VERANSTALTUNGEN** geeignet ist. Das Foyer erhält einen rückwärtigen Ausgang zum attraktiven, grünen Hof des Biologie-Campus.

Die Anlieferung befindet sich auf der Gebäuderückseite im Hof des Biologie-Campus.
ALLE ANLIEFERUNGSBEREICHE, Post- und Paketverteilung, Müllraum, Gasflaschenlager sowie Lösungsmittel- und Chemikalienlager werden in direkter Nachbarschaft zum Lastenaufzug **WIRTSCHAFTLICH GEBÜNDELT.**

Nutzungs- und Zirkulationskonzept

埃森大学附属医院儿科门诊及核医学研究大楼

德国 埃森

2014

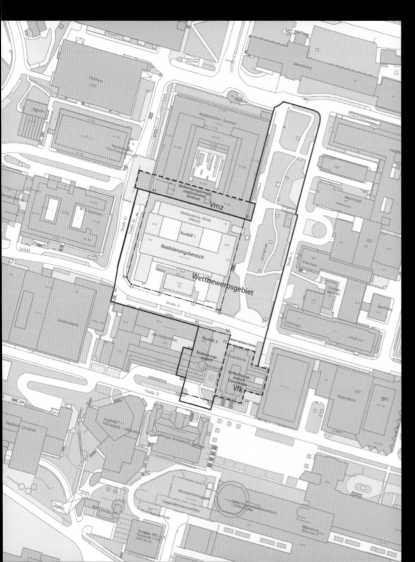

概况

业主：埃森大学附属医院

项目规模：约 11 500 平方米

类型：双阶段开放型设计竞赛（开放申请程序）

参赛者：第一阶段 16 组，第二阶段 6 组

竞赛预算：340 000 欧元（第二阶段奖金 156 000 欧元，第一阶段每组参赛者的费用 7000 欧元，第二阶段每组参赛者的费用 12 000 欧元）

评委

Hans-Jürgen Best, City Director, Planning Unit, Essen; Juliane Brauner, Project Sponsor; Prof. Dr. Ursula Felderhoff-Müser, Director of Pediatric Clinic I, Essen University Hospital; Prof. Dr. M. Norbert Fisch, Technical Building Services Engineer, Braunschweig; Prof. Gesche Grabenhorst, Architect, Hanover; Prof. Linus Hofrichter, Architect, Ludwigshafen; Prof. Dr. Peter F. Hoyer, Director of Pediatric Clinic II, Essen University Hospital; Heiko Schiller, Technical Building Services Engineer, Hamburg; Stephan Triphaus, Architect, UKM Infrastruktur Management GmbH, University Hospital Muenster (UKM)

"特别是对于危重患儿，人性关怀对其治疗和康复尤为重要。新建筑的设计理念应强调并体现为患者提供关爱，并为其创造最佳的治疗和康复条件，切实贯彻'以人为本'，因为医疗进步不仅仅依赖于先进的医疗技术，还与健康、宜人的治疗和康复环境密切相关。"

——引自《竞赛摘要》

埃森大学附属医院

埃森大学附属医院位于鲁尔工业区城市群中，是德国境内的大学附属医院中规模最大的护理中心所在地和尖端医学研究的先导者。医院每年接待15万名门诊病人和5万名住院病人。各科室用房分散于整个校园内，这种布局方式不利于空间功能的协调组织。2000—2010年战略规划初步构想了结构多样且功能完备的校园，其核心是新医学中心大楼，于2011年一期建成，用于药物临床研究。

竞赛场地

儿科门诊及核医学研究大楼位于校园南北轴线的人行通道上，紧临新医学中心。建筑的一部分位于道路另一侧，是儿科门诊和妇女体检中心的所在地。总体而言，场地特征是醒目的山坡地形和较高的建筑密度，建筑技术及功能形象的展示占主导地位。

竞赛任务

儿科门诊及核医学研究大楼占地面积为11 500平方米。该建筑满足了现代临床管理与沟通、教学与科研的所有要求。基地西部和南部的建筑和绿化轴线为校园提供了秩序和方向。同在一栋建筑内的儿科门诊和核医学研究部门因要求连接病房和临近建筑的其他科室，复杂性进一步增加。医院施工和维修过程中的空间定位借助于附近的直升机停机坪。

埃森大学附属医院儿科门诊及核医学研究大楼，埃森

第一阶段参赛者

第二阶段入围者

1 Brunet Saunier Architecture, Paris (F), with Kemper Steiner & Partner Architekten + Stadtplaner, Bochum (D) 2 gmp International, Aachen (D) 3 Heinle, Wischer Gesellschaft für Generalplanung, Berlin (D) 4 HWP Planungsgesellschaft, Stuttgart (D) 5 Ludes Generalplaner, Berlin (D) 6 Nickl & Partner Architekten, Munich (D)

其他参赛者

7 AEP Architekten Eggert Generalplaner, Stuttgart (D) 8 ATP Architekten und Ingenieure, Innsbruck (A) 9 Baumschlager Eberle Gruppe | BE Berlin, Berlin (D) 10 Brechensbauer Weinhart + Partner Architekten, Munich (D) 11 Hascher Jehle Planen und Beraten, Berlin (D), with Monnerjan · Kast · Walter Architekten, Düsseldorf (D) 12 HDR TMK Planungsgesellschaft, Düsseldorf (D) 13 LUDES Architekten – Ingenieure, Recklinghausen (D) 14 pbr Planungsbüro Rohling Architekten und Ingenieure, Braunschweig (D), with HSP Hoppe Sommer Planungsgesellschaft, Stuttgart (D) 15 Schuster Pechtold Schmidt Architekten, Munich (D) 16 tönies+schroeter+jansen freie architekten, Lübeck (D)

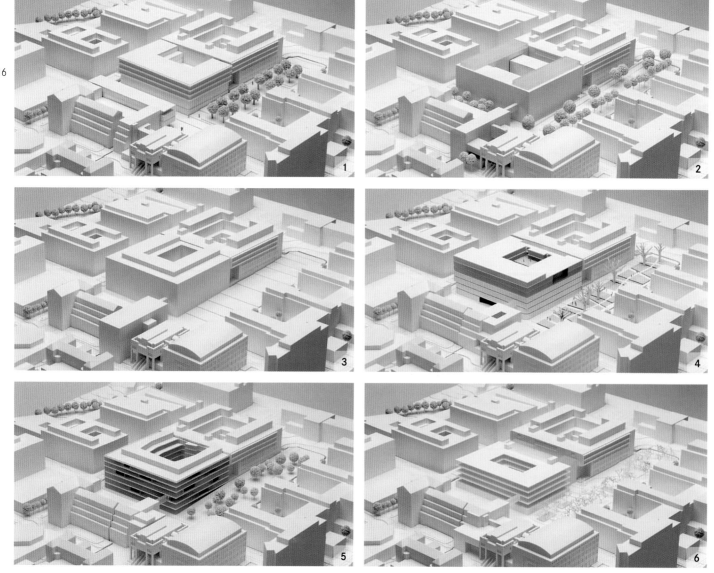

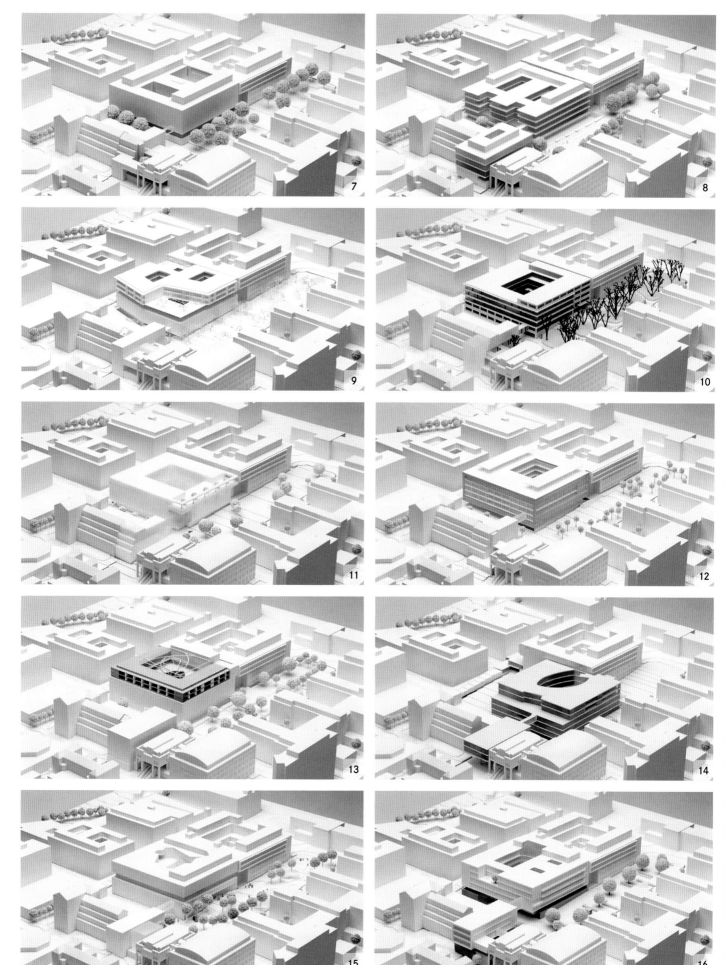

第二阶段参赛者

获奖者
1 一等奖 Heinle, Wischer Gesellschaft für Generalplanung, Berlin (D)
2 二等奖 Ludes Generalplaner, Berlin (D)
3 三等奖 Nickl & Partner Architekten, Munich (D)
4 四等奖 Brunet Saunier Architecture, Paris (F), with Kemper Steiner & Partner Architekten + Stadtplaner, Bochum (D)
5 第二轮 gmp International, Aachen (D)
6 第二轮 HWP Planungsgesellschaft, Stuttgart (D)

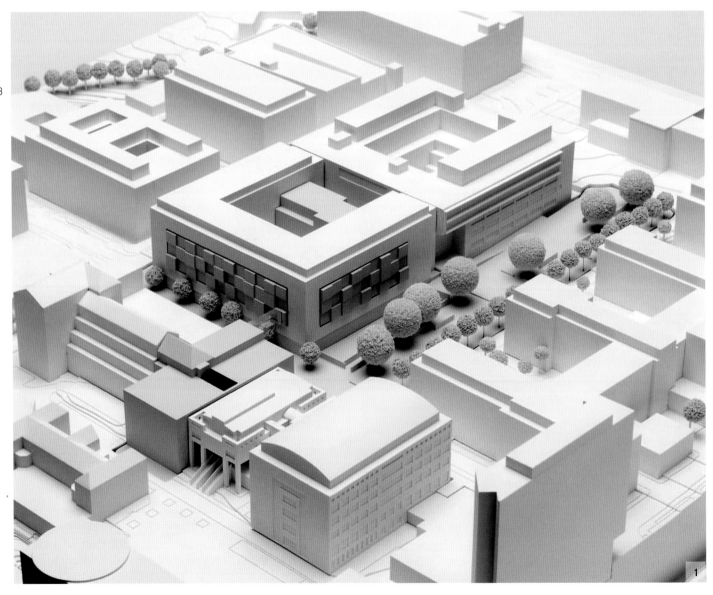

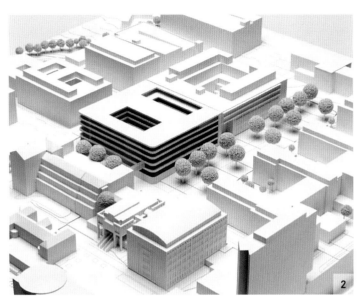

2

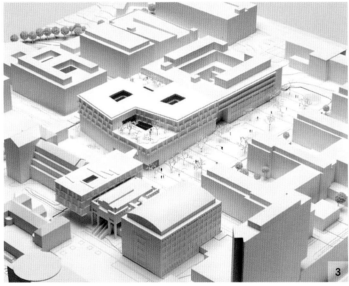

3

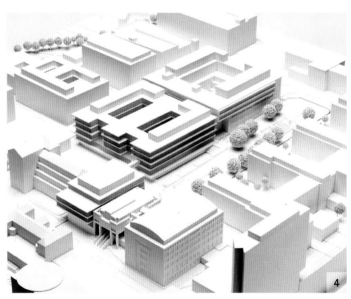

4

5

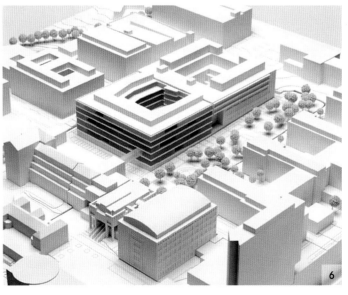

6

竞赛一等奖　获奖事务所

Heinle Wischer
Gesellschaft für Generalplanung

Berlin (D); with Rentschler und Riedesser Ingenieurgesellschaft, Berlin (D);
Wetzel & von Seht, Hamburg (D); RMP Stephan Lenzen Landschaftsarchitekten, Bonn (D)

方案设计者： Edzard Schultz Employees Arianna Bonfatti, Jan Giesen, Hana Michálková, Gautam Shastri
其他顾问： IFG - Ingenieurbüro für Gesundheitswesen, Leipzig (D); WSGreenTechnologies, Stuttgart (D)

  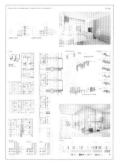 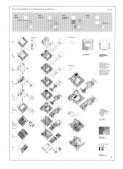 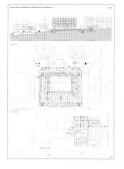 
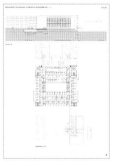 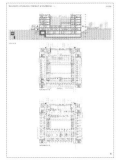   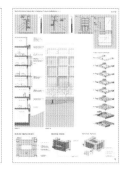 

Heinle Wischer Gesellschaft für Generalplanung

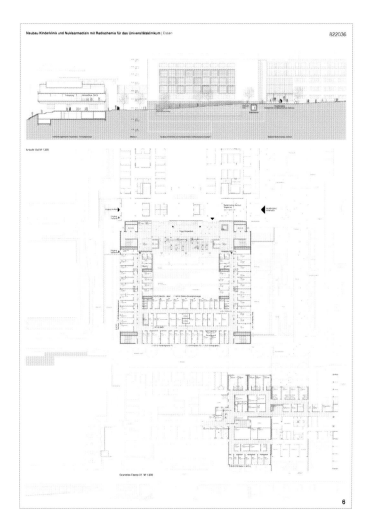 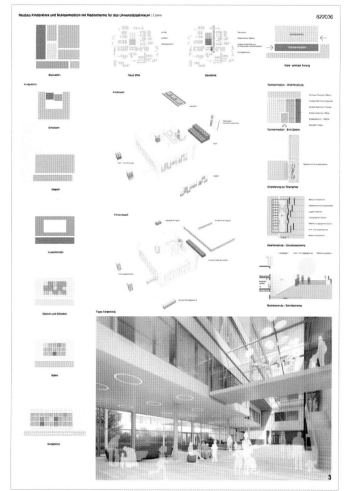

埃森大学附属医院儿科门诊及核医学研究大楼，埃森

竞赛二等奖　获奖事务所

Ludes Generalplaner

Berlin (D); with N.I.L. Ingenieurgesellschaft, Berlin (D); Ingenieurbüro Horn + Horn, Neumünster (D); Rainer Schmidt Landschaftsarchitekten, Berlin (D)

方案设计者：Stefan Ludes
参与者：Sabina Grote-Schepers, David Hupfer, Wojtek Kaminsky, Lydia Kittelmann, Christina Krüger, Marco Neumann

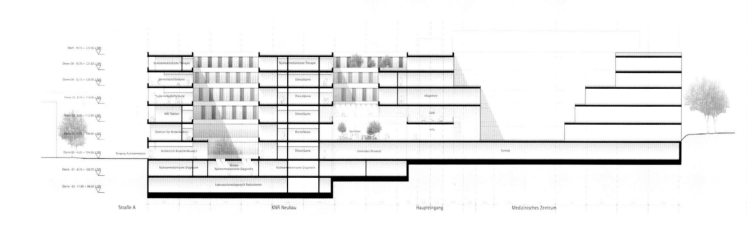

埃森大学附属医院儿科门诊及核医学研究大楼，埃森

竞赛三等奖　获奖事务所

Nickl & Partner Architekten

Munich (D); with Süss Beratende Ingenieure, Nuremberg (D);
Mathes Beratende Ingenieure, Chemnitz (D);
Latz + Partner LandschaftsArchitekten Stadtplaner, Kranzberg (D)

方案设计者：Christine Nickl-Weller
参与者：Daniel Güthler, Nadine Koch, Dragana Simeunovic, Johanna Stemper, Patrik Uchal

Grundrissorganisation

Eingangshalle

Ausblick von der Magistrale

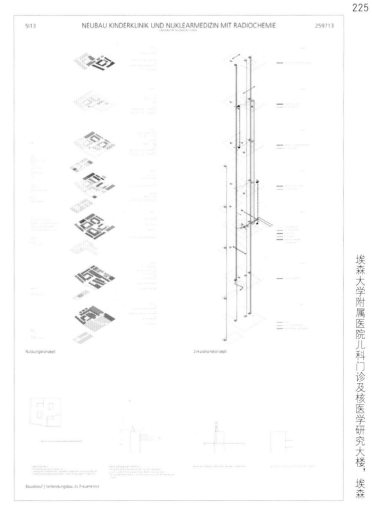

竞赛四等奖 获奖事务所

Brunet Saunier Architecture

Paris (F); with Kemper Steiner & Partner Architekten + Stadtplaner, Bochum (D); ISW – Ingenieur Schmidt & Willmes, Hamm/GFK Gesellschaft für Krankenhausberatung, Cologne (D); Krätzig & Partner Ingenieurgesellschaft für Bautechnik, Bochum (D); SYMplan Landschaftsarchitekturbüro, Essen (D)

方案设计者：Jér.me Brunet, Astrid Beem, Rainer Kemper
参与者：Caspar Muschalek, Catherine Gillier, Magda Sroczynska, Mounia Sa.a
自由职业者：Gerold Zimmerli

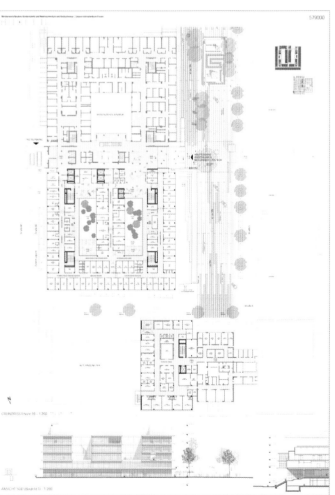
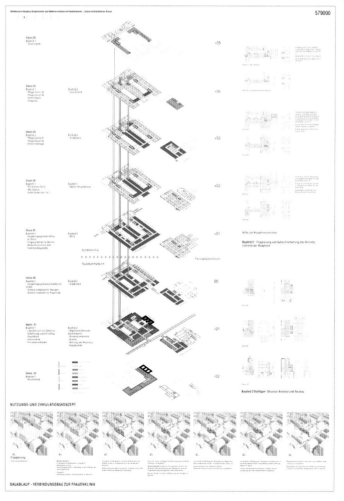

Brunet Saunier Architecture

埃森大学附属医院儿科门诊及核医学研究大楼，埃森

联邦警察总部大楼

德国 波茨坦

2013—2014

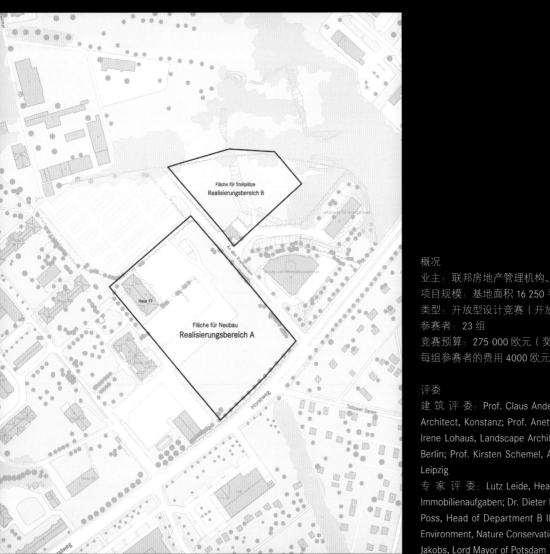

概况

业主：联邦房地产管理机构、勃兰登堡州房地产开发和建设办公室
项目规模：基地面积 16 250 平方米
类型：开放型设计竞赛（开放申请程序）
参赛者：23 组
竞赛预算：275 000 欧元（奖金 140 000 欧元，提名奖 35 000 欧元，每组参赛者的费用 4000 欧元）

评委

建筑评委：Prof. Claus Anderhalten, Architect, Berlin; Prof. Lydia Haack, Architect, Konstanz; Prof. Anett-Maud Joppien, Architect, Darmstadt; Prof. Irene Lohaus, Landscape Architect, Hanover; Elisabeth Rüthnick, Architect, Berlin; Prof. Kirsten Schemel, Architect, Berlin; Benedikt Schulz, Architect, Leipzig

专家评委：Lutz Leide, Head of Facility Management, Bundesanstalt für Immobilienaufgaben; Dr. Dieter Romann, President of the Federal Police; Ralf Poss, Head of Department B II-Federal Buildings, Federal Ministry for the Environment, Nature Conservation, Building and Nuclear Safety (BMUB); Jann Jakobs, Lord Mayor of Potsdam

"竞赛目标是建造一座面向未来的建筑，体现既创新又经济的运营模式，符合必要的安全标准，营造舒适、宜人的工作环境，践行可持续发展的建筑理念。"

——引自《竞赛摘要》

背景信息
联邦警察总部成立于 2008 年，是一个监督与协调联邦警察部队的组织机构。在竞赛期间，新成立的组织机构被暂时安置在波茨坦的三个不同地点。联邦警察执行来自德国国内和国际的具体任务，指令均来自指挥部。

场地位置
新联邦警察总部大楼位于波茨坦海因里希 - 曼 - 阿利 103 号地块，其他州的相关部门也位于此。目前，波茨坦的城市规划可追溯到 Wilhelmstift 建筑时期；形成于 19 世纪下半叶的彩色粉刷外墙和红色砖底座至今仍是基地建筑的主要特征。竞赛场地包括两大地块：一块供新建筑使用，并整合现有建筑；另一块用于停车设施的建设。大街上的"An den Kopfweiden"位于两大地块之间。

竞赛任务
竞赛对指挥控制中心各安全领域的规划质量提出了很高的要求。竞赛任务是建造一片具有代表性的建筑组群，面积 15 000 平方米，可容纳近 1000 名员工。建筑组群应符合德国有关部门和各地工程设计标准，并为未来的相关建设作出示范。工程重点在于高质量的建筑内部组织，特别是为指挥控制中心以及办公区提供最佳设计。

联邦警察总部大楼，波茨坦

参赛者

获奖者

1 一等奖 CODE UNIQUE, Dresden (D), with herrburg Landschaftsarchitekten, Berlin (D)
2 二等奖 gmp International with Man Made Land, Berlin (D)
3 三等奖 Bodamer Faber Architekten with Plankontor S1 Landschaftsarchitekten, Stuttgart (D)
4 荣誉奖 hks Hestermann Rommel Architekten with plandrei Landschaftsarchitektur, Erfurt (D)
5 荣誉奖 h4a Gessert + Randecker + Legner Architekten, Stuttgart (D), with Planstatt Senner, .berlingen (D)
6 决赛者 Weinmiller Architekten with LA.BAR Landschaftsarchitekten, Berlin (D)

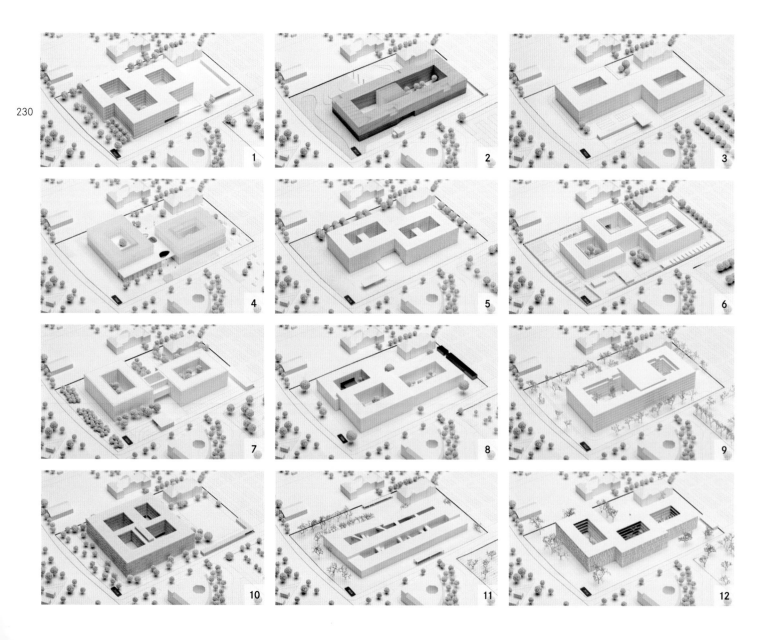

第二轮

7 Thomas Müller Ivan Reimann Architekten with Vogt Landschaftsarchitekten, Berlin (D) **8** BLK2 Böge Lindner K2 Architekten with Schoppe + Partner Freiraumplanung, Hamburg (D) **9** Klein und Sänger Architekten, Munich (D), with Claudia Weber-Molenaar, Gräfelfing (D) **10** BHBVT Architekten with Stefan Bernard Landschaftsarchitekten, Berlin (D) **11** Baumschlager Eberle Gruppe | BE Berlin with bbz Landschaftsarchitekten Berlin, Berlin (D) **12** Numrich Albrecht Klumpp Architekten, Berlin (D), with Marcel Adam Landschaftsarchitekten, Potsdam (D) **13** Gatermann + Schossig Bauplanungsgesellschaft, Cologne (D), with Fugmann Janotta Landschaftsarchitektur und Landschaftsentwicklung, Berlin (D) **14** thoma architekten + Kummer.Lubk.Partner, Berlin/Erfurt (D), with Heinisch Landschaftsarchitekten, Gotha (D) **15** TREUSCH architecture with Idealice – technisches büro für landschaftsarchitektur, Vienna (A) **16** Bruno Fioretti Marquez Architekten with capatti staubach Urbane Landschaften, Berlin (D) **17** AllesWirdGut Architektur, Vienna (A), with t17 Landschaftsarchitekten, Munich (D)

第一轮

18 Hascher Jehle Planungsgesellschaft with Weidinger Landschaftsarchitekten, Berlin (D) **19** Heinle, Wischer und Partner Freie Architekten with Hager Partner, Berlin (D) **20** Nickl & Partner Architekten, Munich (D), with GHP Landschaftsarchitekten, Hamburg (D) **21** KSV Krüger Schuberth Vandreike with Hanke + Partner Landschaftsarchitekten, Marek Jahnke, Berlin (D) **22** Atelier ST Architekten with Matthias Lanzendorf Landschaftsarchitekten, Leipzig (D) **23** wulf architekten with Jetter Landschaftsarchitekten, Stuttgart (D)

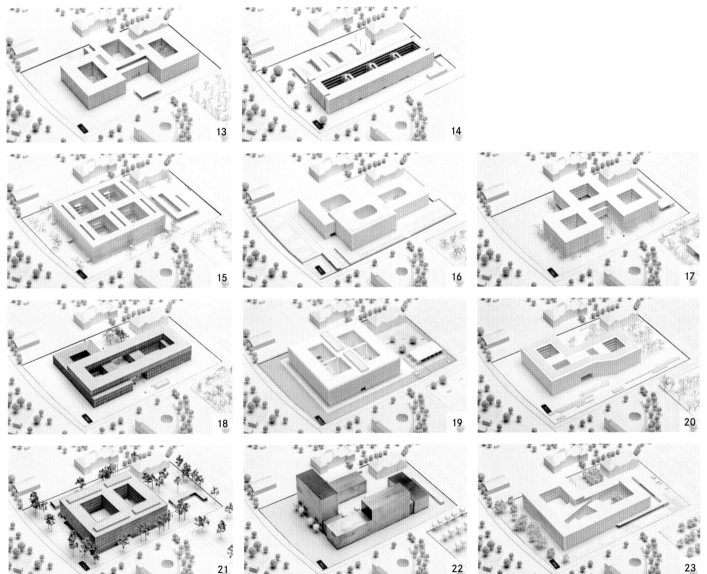

竞赛一等奖　获奖事务所

CODE UNIQUE

Dresden (D); with herrburg Landschaftsarchitekten, Berlin (D)

方案设计者：Volker Giezek, Martin Boden-Peroche, Mareike Sch.nherr
参与者：Peter Weber, Claire Dupré, Luisa G.hlert, Michael Klemm, Peter Jarisch

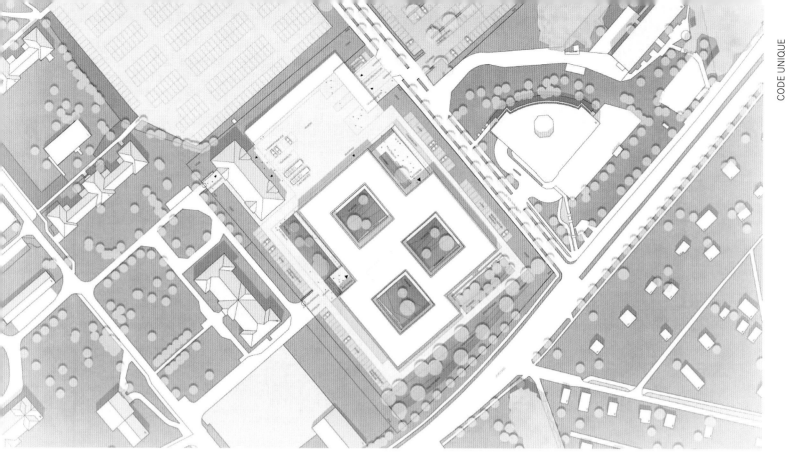

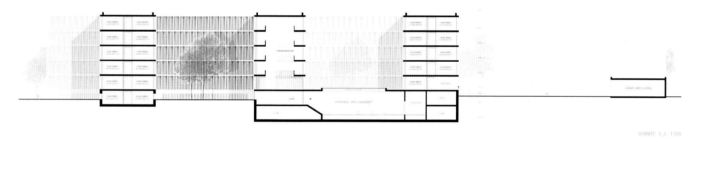

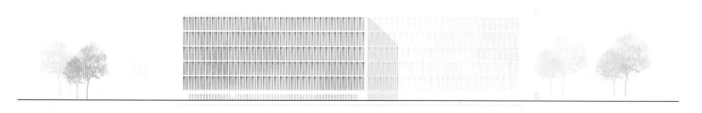

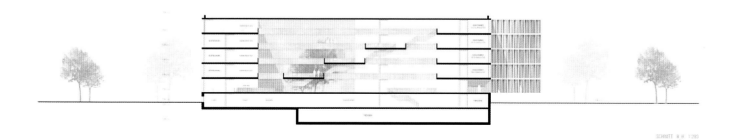

CODE UNIQUE

联邦警察总部大楼，波茨坦

竞赛二等奖　获奖事务所

gmp International

Berlin (D); with Man Made Land, Berlin (D)

方案设计者： Hubert Nienhoff, Alexandre Mellier
参与者： Martin Glass, Christiane Wermers, Anne Gross, Florian Alles, Camila Preve, Ana Tendeiro, Elke Glass, Claudia Chiappini, Fossati Alessio, Christian Bohne, Massimo de la Rosa
顾问： Transsolar Energietechnik, Stuttgart (D); Hartwich/Mertens/Ingenieure Planungsgesellschaft für Bauwesen, Berlin (D)

140392

gmp International

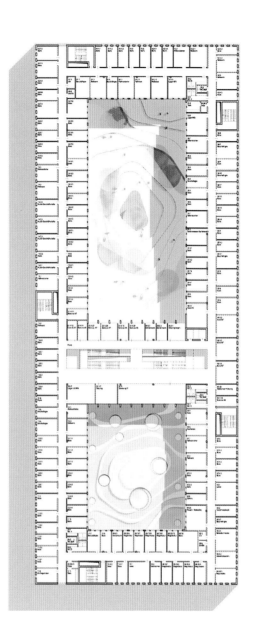

3. Obergeschoss M 1:200

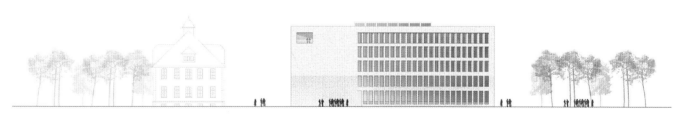

Ansichten Süd-Ost und Süd-West M 1:200

Neubau Bundespolizeipräsidium in Postdam

联邦警察总部大楼，波茨坦

Blatt 06

竞赛三等奖　获奖事务所

Bodamer Faber Architekten

Stuttgart (D); with Plankontor S1 Landschaftsarchitekten, Stuttgart (D)

方案设计者： Hansj.rg Bodamer, Ulrich Schuster
参与者： Daniel Heyer, Johanna Borrmann, Sandra Golinski, Antje Hipp
顾问： WSGreenTechnologies, Stuttgart (D)

Neubau Bundespolizeipräsidium in Potsdam

Grundriss Erdgeschoss und Freianlagen 1:200

席勒公园房地产开发

德国 柏林

2012

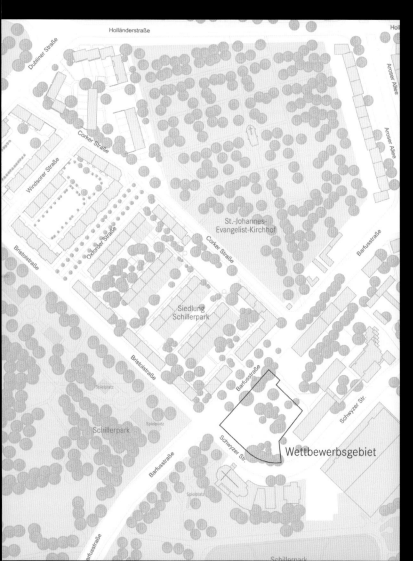

概况

业主:1892 柏林建设和住房合作社

项目规模:约 90 个居住单元

类型:设计竞赛

参赛者:10 组

竞赛预算:54 400 欧元(奖金 12 400 欧元,每组参赛者的费用 3500 欧元)

评委

建筑评委:Andreas Becher, Architect, Berlin; Prof. Max Dudler, Architect, Berlin/Zurich; Prof. Gisela Glass, Architect, Berlin; Muck Petzet, Architect, Munich; Regina Jost, Architect, Berlin

专家评委:Manfred Kühne, Head of Department, Senate Department for Urban Development and the Environment, Dept. II, Berlin; Kristina Laduch, Head of City Planning Office, District Berlin-Mitte; Thorsten Schmitt, Board Member, Berliner Bau- und Wohnungsgenossenschaft von 1892 eG, Berlin; Monika Markgraf, Bauhaus Dessau Foundation, International Council on Monuments and Sites (ICOMOS) Monitoring; Dirk Lönnecker, Board Member, Berliner Bau- und Wohnungsgenossenschaft von 1892 eG, Berlin; Alois Wortmann, Department of City Planning, Planning and Building Permission Office, District Berlin-Mitte; Dr. Dagmar Tille, Head of Building Culture, Communication, Historic Preservation Authority Workshop, Senate Department for Urban Development and the Environment Berlin

"竞赛场地位于联合国教科文组织世界文化遗产地的部分区域中,在现有历史建筑的缓冲区内。保护席勒公园定居点为方案设计提出了特别的挑战。"

——引自《竞赛摘要》

1892 柏林建设和住房合作社

1892 柏林建设和住房合作社(以下简称"1892 合作社")是柏林市第二大建设和住房合作社,它是在住房短缺和物业不完备的情况下建立的,从成立伊始即奉行改良主义的建筑理念(影响着建筑甚至社会文化的方方面面)。20 个席勒公园定居点经过多年的建设已日趋成熟。"1892 合作社"中最具影响力的建筑师以艾尔弗雷德梅塞尔、布鲁诺陶特和汉斯·霍夫曼为代表。这些定居点均建立了适宜的规划框架,以舒适、可持续和高品质作为根本出发点。2012 年是"1892 合作社"成立 120 周年,这一年被联合国定为"国际合作年",也是席勒公园房地产开发竞赛举办之年。

席勒公园定居点

该地块位于德国柏林 Wedding 区 Schwyzer 大街上,是 Barfuss 街区的一个角落。位于南部的席勒公园和始建于 1956 年的圣阿洛伊修斯教堂是该地区的重要地标。面向西北的席勒公园定居点作为 1924 年初一战之后的第一个社会住宅建设项目历经了三个建设阶段。在 1954—1959 年间,定居点的范围不断扩大,现在已有约 600 套公寓。2008 年 7 月,定居点与其他五座德国古典主义现代风格公寓一同被联合国教科文组织列入"世界遗产名录新增项目"。

竞赛任务

基于 2010 年通过的发展计划,公寓包含地下停车场和至少 5200 平方米的出租面积。竞赛任务的一个重点是解决现有定居点规划和建筑方面的问题,并使相关问题的"转译"体现在建筑平面布局、户外公共空间和绿化带中。公寓中,除了一居室、二居室、三居室和四居室,还包括可供 10 位老年人居住的养老公寓。

席勒公园房地产开发,柏林

获奖者

1 一等奖 Bruno Fioretti Marquez Architekten, Berlin (D)
2 二等奖 HAAS Architekten, Berlin (D)
3 三等奖 blauraum architekten, Hamburg (D)

参赛者

第二轮

4　hildebrandt.lay.architekten, Berlin (D)
5　Fink + Jocher Architekten und Stadtplaner, Munich (D)
6　Mola + Winkelmüller Architekten, Berlin (D)

第一轮

7　zanderroth architekten, Berlin (D)
8　Léon Wohlhage Wernik Architekten, Berlin (D)
9　Baumschlager Eberle Gruppe | BE Berlin, Berlin (D)
10　Winfried Brenne Architekten, Berlin (D)

竞赛一等奖　获奖事务所

Bruno Fioretti Marquez Architekten

Berlin (D)

方案设计者：Piero Bruno
参与者：Michael Flenske, Simon Filler, Lorenz Kirchner, Paul Künzel, Latittia Vouillon, Julia Henning
顾问：ifb frohloff staffa kühl ecker, Berlin (D), Henning Ecker; elephantgreen cgi, Berlin (D), Max Nallenweg; Vogt Landschaftsarchitekten, Zürich (CH); IG Zimmermann, Berlin (D)

Wohnungsneubau an der "Siedlung Schillerpark"

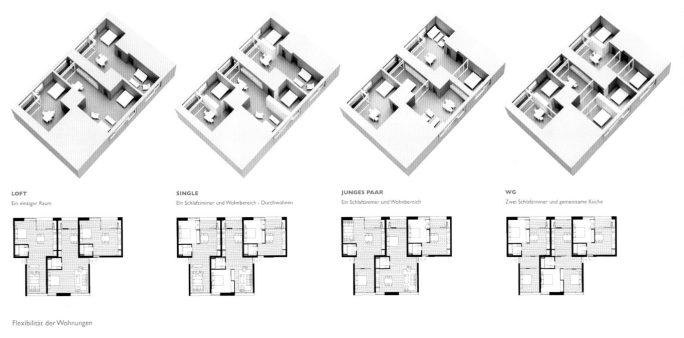

LOFT
Ein einziger Raum

SINGLE
Ein Schlafzimmer und Wohnbereich - Durchwohnen

JUNGES PAAR
Ein Schlafzimmer und Wohnbereich

WG
Zwei Schlafzimmer und gemeinsame Küche

Flexibilität der Wohnungen

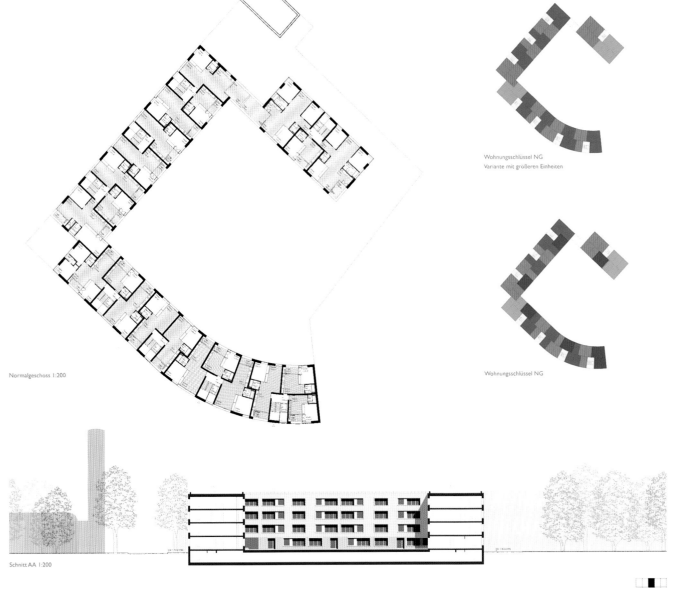

Normalgeschoss 1:200

Wohnungsschlüssel NG
Variante mit größeren Einheiten

Wohnungsschlüssel NG

Schnitt AA 1:200

花园城区法尔肯堡居住区

德国 柏林

2014

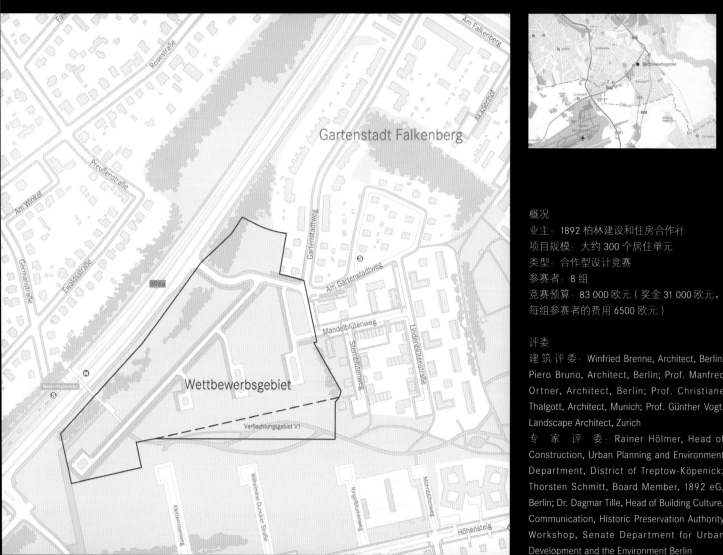

概况
业主：1892 柏林建设和住房合作社
项目规模：大约 300 个居住单元
类型：合作型设计竞赛
参赛者：8 组
竞赛预算：83 000 欧元（奖金 31 000 欧元，每组参赛者的费用 6500 欧元）

评委
建筑评委：Winfried Brenne, Architect, Berlin; Piero Bruno, Architect, Berlin; Prof. Manfred Ortner, Architect, Berlin; Prof. Christiane Thalgott, Architect, Munich; Prof. Günther Vogt, Landscape Architect, Zurich
专家评委：Rainer Hölmer, Head of Construction, Urban Planning and Environment Department, District of Treptow-Köpenick; Thorsten Schmitt, Board Member, 1892 eG, Berlin; Dr. Dagmar Tille, Head of Building Culture, Communication, Historic Preservation Authority Workshop, Senate Department for Urban Development and the Environment Berlin

"竞赛任务的一个重点是协调项目基地在世界历史文化遗址保护与城市规划及建筑设计两方面的关系，将历史设想中的创新型住宅融入现代化的建筑平面布局、户外公共空间和绿化带中。"

——引自《竞赛摘要》

1892柏林建设和住房合作社

在组织举办了席勒公园房地产开发竞赛之后，"1892合作社"又组织举办了另一个房地产开发项目的两场竞赛，设计对象是在2008年与其他五座德国古典主义现代风格公寓一同被联合国教科文组织列入"世界遗产名录新增项目"的花园城区法尔肯堡居住区（绰号"颜料盒定居点"）。

花园城区法尔肯堡居住区

花园城区法尔肯堡居住区坐落在德国柏林东南部特雷普托克佩尼克区。在1913—1916年间，居住区由布鲁诺陶特和路德维希来瑟尔设计，自2002年经历了三次扩建。至2014年末，居住区共包含53座住宅和342套公寓。竞赛选址位于世界遗产地的缓冲区。项目用地边界的96a公路和与街道平行的有7米高差的堤坝是场地规划主要影响因素。场地南部边界处的公园将开发为中央休闲空间，服务于周边的居住综合体。

竞赛任务

竞赛任务是打造一片至少28 000平方米的居住区，根据城市总体规划的初稿对各种类型的建筑进行集成式设计，包括优化城市结构规划和建筑及室外设施设计。房屋应该供两代人使用，是实用、高效且平均净租金低于每平方米10欧元的共享公寓。房屋的特殊性应该体现为结构和品质的高度融合，包括室内空间和室外空间。根据"花园城区"的规划理念，竞赛任务的重中之重是室外空间一体化设计以及公私区域的合理布局。

花园城区法尔肯堡居住区，柏林

参赛者

获奖者

1 一等奖 zanderroth architekten, Berlin (D)
2 二等奖 ROBERTNEUN ™ Architekten, Berlin (D)
3 三等奖 Heidenreich & Springer Architekten, Berlin (D)

第三轮

4 Blumers Architekten Generalplanung und Baumanagement, Berlin (D)

第二轮

5 Heide & von Beckerath Architekten, Berlin (D)

第一轮

6 HAAS Architekten, Berlin (D)
7 Thomas Müller Ivan Reimann Architekten, Berlin (D)
8 brh Architekten + Ingenieure, Berlin (D)

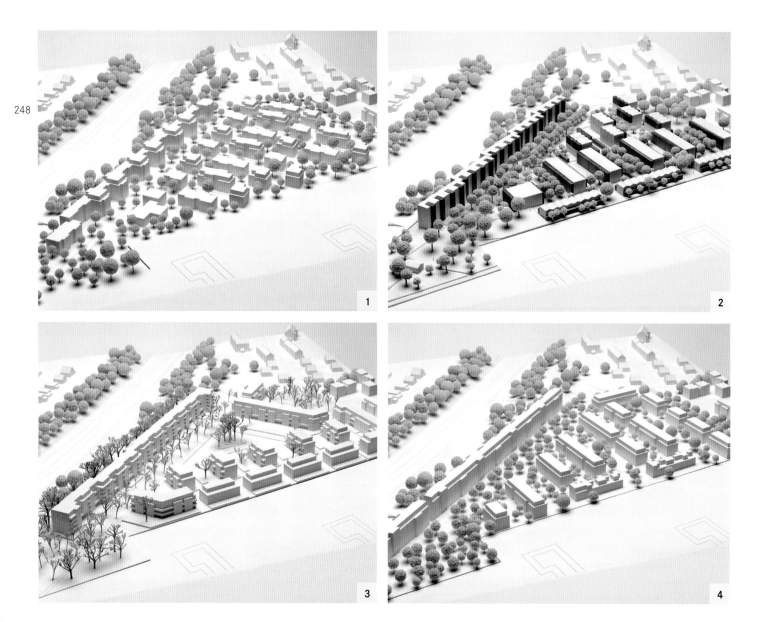

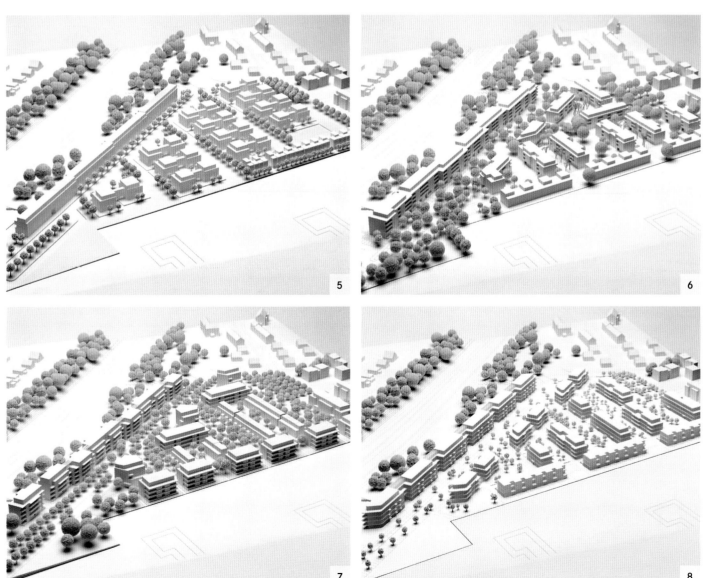

5
6
7
8

竞赛一等奖　获奖事务所

zanderroth architekten

Berlin (D)

方案设计者：Sascha Zander, Christian Roth
参与者：Sofia Vaasen, Nils Schülke, Alexander Markau, Sebastian Kern, Martin Tessarz
顾问：Hager Partner, Landschaftsarchitekten, Berlin (D); Zimmermann und Becker, Berlin (D); Insar | schwartze, wessling und partner, Berlin (D)

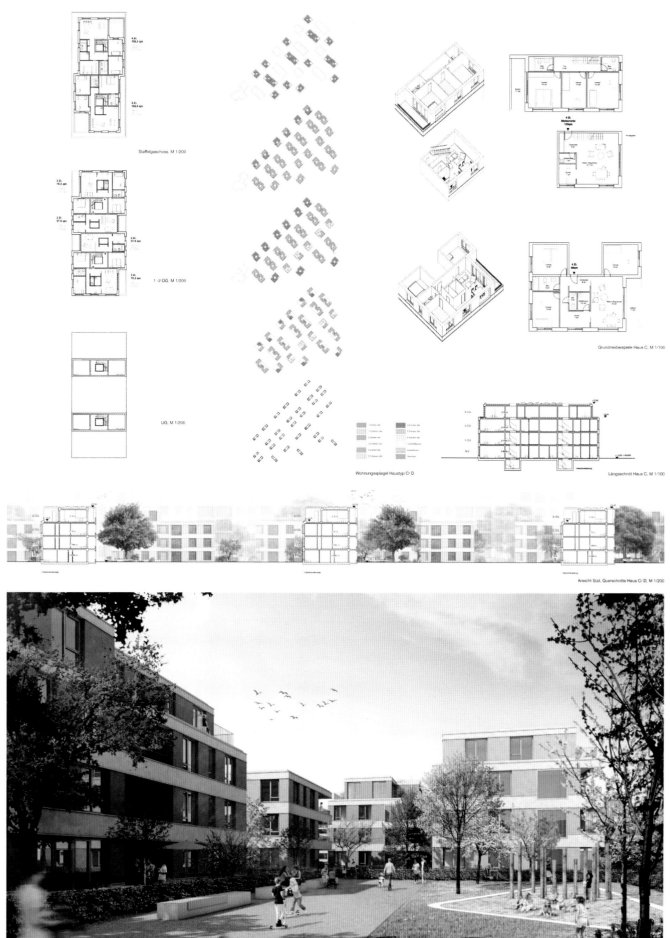

竞赛二等奖　获奖事务所

ROBERTNEUN™ Architekten

Berlin (D)

方案设计者： Nils Buschmann, Tom Friedrich
参与者： Christoph Michael, Theresa Behling, Nina Dvorak
顾问： Atelier Loidl Landschaftsarchitekten, Berlin (D), Leo Grosch; BAI Building Applications Ingenieure - Kasche Lu.ky Dr. Krühne, Berlin (D)

竞赛二等奖　获奖事务所

Heidenreich & Springer Architekten

Berlin (D)

方案设计者：J.rg Springer
参与者：Max Wasserkampf, Tornike Kublashvili, Fabian Lux, Julia Naumann
顾问：Stefan Bernard Landschaftsarchitekten, Berlin (D)

Ansicht Süd M 1:200

Ansicht Nord M 1:200

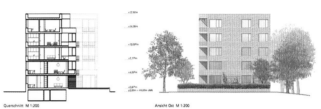
Querschnitt M 1:200   Ansicht Ost M 1:200

4 Zimmer Wohnung M 1:100

Wohngruppe M 1:100

Staffelgeschoss M 1:200

2./3. Obergeschoss M 1:200

1. Obergeschoss M 1:200

Erdgeschoss M 1:200

Wohnungsschlüssel Erdgeschoss

Wohnungsschlüssel 1. Obergeschoss

Wohnungsschlüssel 2./3. Obergeschoss

Wohnungsschlüssel Staffelgeschoss

下萨克森州议会大厦

德国 汉诺威

2009—2010

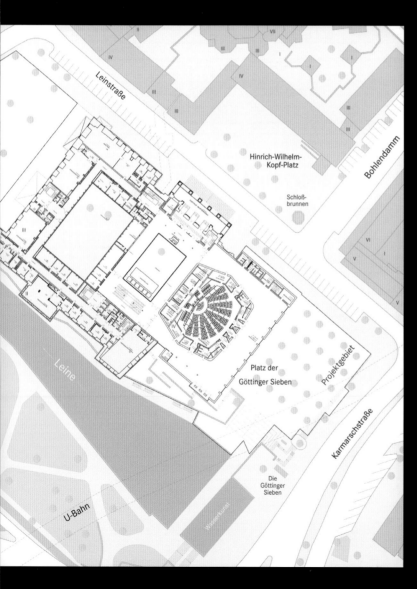

概况
业主：下萨克森州，汉诺威联邦政府公共工程管理局
项目规模：基地面积 8500 平方米
类型：双阶段设计竞赛
参赛者：第一阶段 57 组，第二阶段 16 组
竞赛预算：171 000 欧元（奖金 137 000 欧元，荣誉奖 34 000 欧元）

评委
建筑评委：Prof. Carl Fingerhuth, Architect and City Planner, Zurich/Darmstadt; Hon. Prof. Johanne Nalbach, Architect, Berlin; Prof. Gernot Schulz, Architect, Cologne; Prof. Arno Lederer, Architect, Stuttgart; Kaspar Kraemer, Architect, Cologne; Prof. Zvonko Turkali, Architect, Frankfurt am Main; Wolfgang Schneider, President of the Chamber of Architects, Lower Saxony; Doris Nordmann, Department Head, Ministry of Finance, Lower Saxony; Uwe Bodemann, Building Director, Hanover

专家评委：Hermann Dinkla, President of the Parliament of Lower Saxony; Astrid Vockert, Vice President of the Parliament of Lower Saxony (CDU); Björn Thümler, Parliamentary Manager, CDU Parliamen-tary Group; Hans-Werner Schwarz, Vice President of the Parliament of Lower Saxony (FDP); Heiner Bartling, Parliamentary Manager, SPD Parliamentary Group; Dr. Silke Lesemann, Parliament Member (SPD); Enno Hagenah, Parliament Member (Bündnis 90/DIE GRÜNEN); Christa Reichwaldt, Parliamentary Manager, DIE LINKE Parliamentary Group

"竞赛理念是在建筑功能、图形的可达性和透明性方面标新立异,超越现有建筑的品质及其限制性。"

——引自《竞赛摘要》

莱茵城堡国家议会

下萨克森州议会自 1962 年以来一直坐落在汉诺威市莱茵城堡。莱茵城堡历史悠久,其前身是建于 1291 年的圣方济各修道院(多种族教堂)。之后若干年,莱茵城堡发生了诸多改变,在 1943 年因空袭几乎被完全摧毁。建筑师 Dieter Oesterlen 在 1954 年基于竞赛方案将其作为国会大厦予以重建,会议室尤其体现了其特点,建筑和空间都是"流动"的,议会成员在"流动"的氛围中工作,不受外界影响和干扰。建筑外墙与城堡大门并排设置,象征着历史与现代的交融。

项目推动力

现有会议区已不再符合国家议会的职能要求与议会工作透明化的办公理念。此外,礼堂区域缺少维护设施和无障碍设施。因此,竞赛主办方决定发起以改造、翻新现有会议区为目的的竞赛,具体的规划区域包括入口区、会议厅、办公室、游客接待中心、多媒体区和餐厅。

新国会大厦

竞赛任务是设计一个会议区,其外观形式特色鲜明,并强调该建筑的多元功能。国会大厦是一座纪念碑式建筑,需要谨慎处理,以呼应现有建筑的物质形态。其他竞赛任务包括:消除空间的功能缺陷,重新设置空间功能,保留建筑的纪念碑部分及其他重要节点。

下萨克森州议会大厦,汉诺威

第一阶段参赛者

第二阶段入围者

1 Bahl + Partner Architekten, Hagen (D) 2 Jabusch + Schneider Architekten, Hanover (D) 3 nps tchoban voss, Berlin (D) 4 Böge Lindner Architekten, Hamburg (D) 5 Walter Gebhardt Architekt, Hamburg (D) 6 Fritzen + Müller-Giebeler Architekten, Ahlen (D) 7 Stephan Braunfels Architekten, Berlin (D) 8 Behles & Jochimsen Architekten, Berlin (D) 9 Mijic Architects, Rimini (I) 10 YI ARCHITECTS, Cologne (D) 11 Cornelsen + Seelinger Architekten, Darmstadt (D) 12 Meyer-Wolters & Yeger Architekten, Hamburg (D) 13 Ring Architekten, Munich (D) 14 Karsten K. Krebs Architekten, Hanover (D) 15 mm architekten, Hanover (D) 16 Stankovic Architekten, Berlin (D)

其他参赛者

17 Fritz-Dieter Tollé Architekten, Verden (D) 18 Manfredi Anello, Dublin (IRL) 19 Kunzemann Architekten, Großburgwedel (D) 20 male architekten mit Jakob Timpe, Berlin (D) 21 Magma Architecture, Berlin (D) 22 Architekturbüro Ostermeyer, Hanover (D) 23 spine architects, Hamburg (D) 24 MMZ Architekten, Hanover (D) 25 Lieseberg Architekten, Hanover (D) 26 Poos Isensee Architekten, Hanover (D) 27 Martienssen Architekten, Hanover (D) 28 argeplan, Hanover (D) 29 Jürgen Scharlach Architektur und Stadtplanung, Isernhagen (D) 30 Mola + Winkelmüller Architekten, Berlin (D)

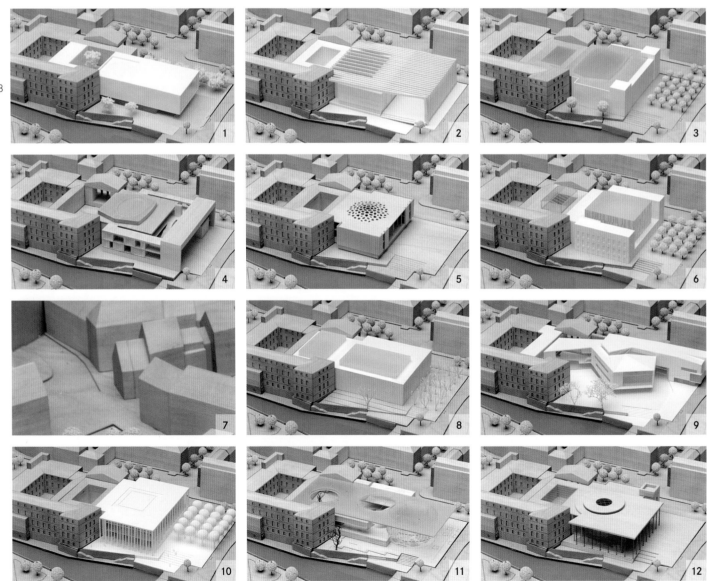

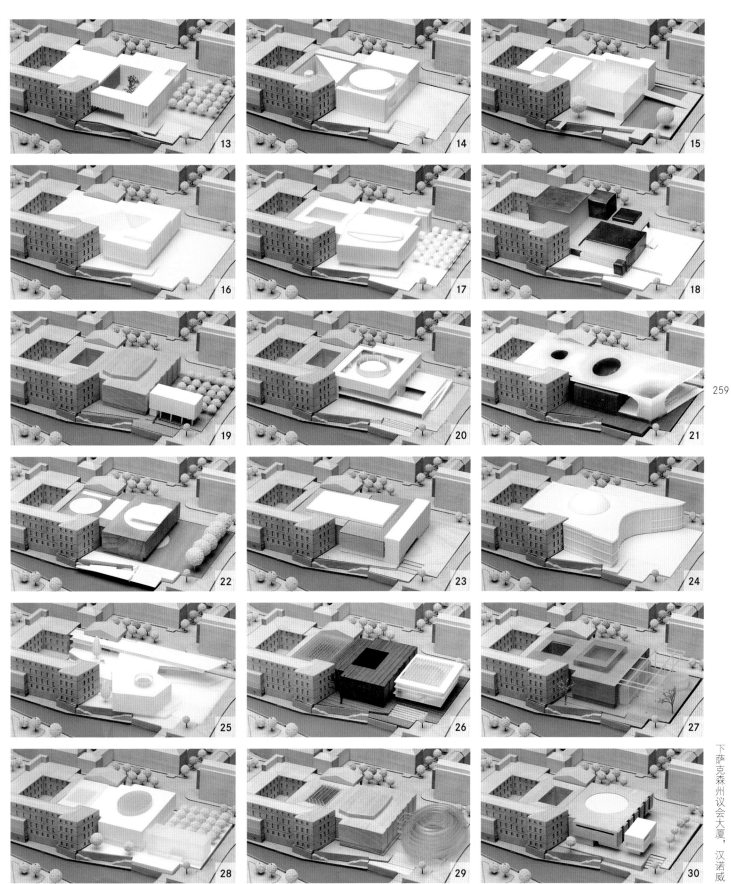

第一阶段其他参与者

31 Nieberg Architekten, Hanover (D) **32** ORG-Design & Architecture, Hanover (D) **33** Meier-Scupin + Partner Architekten, Munich (D) **34** Architekturbüro Törber, Hanover (D) **35** Grüttner Architekten, Soest (D) **36** Architekt Peter Schoof, Hanover (D) **37** SB-Studio, Saarbrücken (D) **38** Prof. Kollhoff Generalplanung, Berlin (D) **39** Nickl & Partner Architekten, Munich (D) **40** Studyo Architects, Cologne (D) **41** Architektur- Stadtplanungsbüro, Westerstede (D) **42** 1010-Architektur, Hanover (D) **43** GMS Freie Architekten, Isny im Allgäu (D) **44** IF architecture, Hanover (D) **45** lüderwaldt architekten, Hanover (D) **46** Drexler Guinand Jausin, Frankfurt am Main (D) **47** Andreas J. Keller, Frankfurt am Main (D) **48** Team Li Sa, Hanover (D) **49** Architekt Backe, Berlin (D) **50** Paul Bretz Architekten, Luxemburg (L) **51** Architekturbüro Lungwitz, Dresden (D) **52** Runge Architekten, Hanover (D) **53** Architekturbüro Professor Wolfgang Kergaßner, Ostfildern (D) **54** Architekten Baumgart Wockenfuß und Partner, Celle (D) **55** Grigat-Architekten, Stadthagen (D) **56** Kölling Architekten, Bad Vilbel (D) **57** Sergio Pascolo Architects, Venice (I)

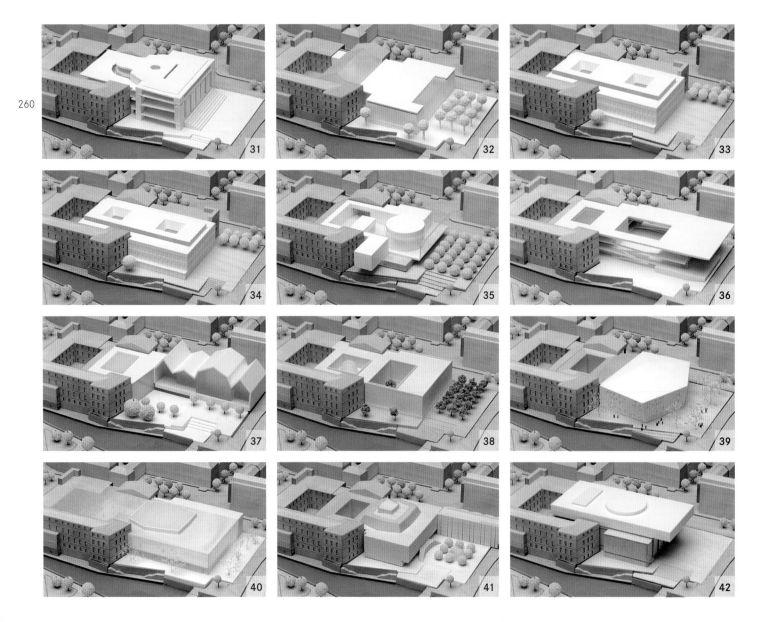

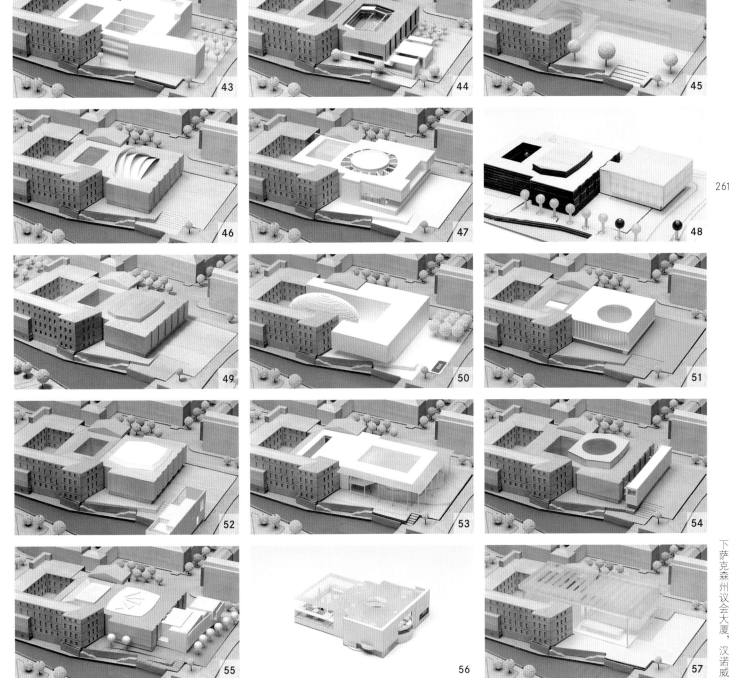

第二阶段参赛者

获奖者

1. 一等奖 YI ARCHITECTS, Cologne (D)
2. 二等奖 Walter Gebhardt Architekt, Hamburg (D)
3. 三等奖 mm architekten, Hanover (D)

第二轮

4. Behles & Jochimsen Architekten, Berlin (D)
5. Ring Architekten, Munich (D)
6. Stephan Braunfels Architekten, Berlin (D)
7. Jabusch + Schneider Architekten, Hanover (D)
8. Meyer-Wolters & Yeger Architekten, Hamburg (D)
9. Fritzen + Müller-Giebeler Architekten, Ahlen (D)

第一轮

10. Bahl + Partner Architekten, Hagen (D)
11. nps tchoban voss, Berlin (D)
12. Cornelsen + Seelinger Architekten, Darmstadt (D)
13. Böge Lindner Architekten, Hamburg (D)
14. Mijic Architects, Rimini (I)
15. Karsten K. Krebs Architekten, Hanover (D)
16. Stankovic Architekten, Berlin (D)

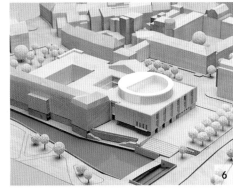
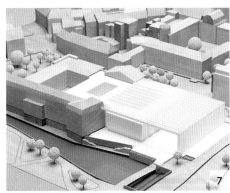

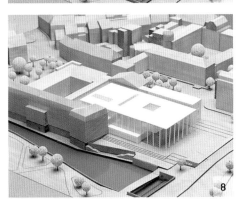

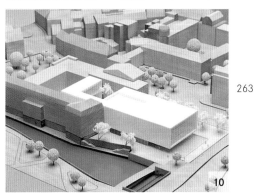

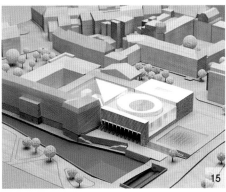
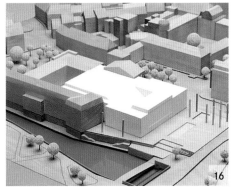

卡尔广场邻近地区

德国 斯图加特

2009

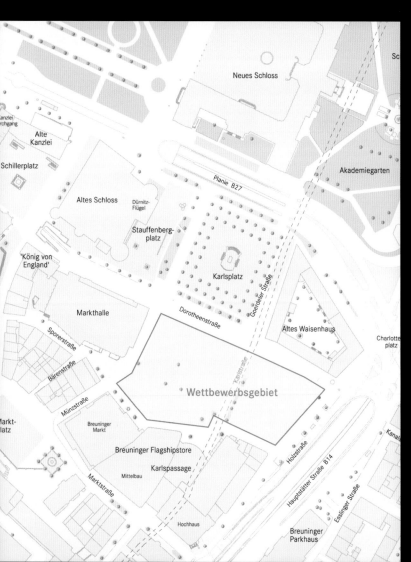

概况

业主：巴登-符腾堡州经济商业局、E. Breuninger 有限公司、EKZ Grundstücksverwaltung 有限公司
项目规模：基地面积 49 000 平方米
类型：开放型设计竞赛（开放申请程序）
参赛者：11 组
竞赛预算：560 000 欧元（奖金 200 000 欧元，费用 30 000 欧元）

评委

建筑评委：Prof. Carl Fingerhuth, Architect, Zurich/Darmstadt; Prof. Marc Angélil, Architect, Zurich/Los Angeles; Prof. Regine Leibinger, Architect, Berlin; Jórunn Ragnarsdóttir, Architect, Stuttgart; Wolfgang Riehle, Architect and City Planner, Reutlingen; Prof. Roger Riewe, Architect, Graz; Prof. Kirsten Schemel, Architect, Berlin/Münster; Kai Fischer, Director of Construction, Baden-Württemberg Ministry of Finance and Economics, Stuttgart; Annette Ipach-Öhmann, Architect, Director of the Baden-Württemberg Administration for Property and Building Construction, Stuttgart; Claudia Reusch, Department Head, Baden-Württemberg Ministry of Finance and Economics, Stuttgart; Frank Dittel, Architect, Stuttgart; Lars Uwe Bleher, Architect, Frankfurt am Main/Portland; Wolf Mizsgar, Architect, Stuttgart

专家评委：Günther H. Oettinger, Member of State Parliament and Prime Minister of the State of Baden-Württemberg; Willi Stächele, Minister of Finance and Economics of the State of Baden-Württemberg; Dr. Gisela Meister-Scheufelen, Department Head, Baden-Württemberg Ministry of Finance and Economics; Willem G. van Agtmael, Chief Executive Officer, E. Breuninger GmbH & Co., Stuttgart; Dr. Wienand Meilicke, Chairman of the Board, E. Breuninger GmbH & Co., Stuttgart; Willy Oergel, Chief Operating Officer, E. Breuninger GmbH & Co., Stuttgart; Dr. Wolfgang Schuster, Mayor of Stuttgart; Matthias Hahn, Deputy Mayor and Head of Department of Urban Development and the Environment, Stuttgart; Philipp Hill, City Councillor, CDU parliamentary group, Stuttgart; Dieter Wahl, City Councillor, CDU parliamentary group, Stuttgart; Manfred Kanzleiter, SPD parliamentary group, Stuttgart; Dr. Michael Kienzle, Bündnis 90/DIE GRÜNEN parliamentary group, Stuttgart

"项目周边社区应该被给予积极、明确的发展推动力，进而成为城镇商业、休闲旅游最活跃的区域。设计应该尊重现有古迹建筑，同时形成有意义的城市规划框架并彰显建筑特色。"

——引自《竞赛摘要》

斯图加特市中心

斯图加特是德国巴登-符腾堡州首府，有6万人口，是德国第六大城市。卡尔广场邻近社区项目是全面重整各州政府相关部门项目的一部分，与各州政府相关部门的重组并行。区域规划包括高品质的零售商店、餐厅和酒店。项目宗旨是增强卡尔广场邻近地区的吸引力并强化其国际定位。项目由巴登-符腾堡州政府以及斯图加特市时装和生活风尚百货公司——E. Breuninger 有限公司共同发起。

卡尔广场

由两个城市街区构成的竞赛区域是建筑林立、人口密集的地区，以斯图加特-米特区的公共空间和历史建筑著称。老城堡与符腾堡州立博物馆对面是新城堡，它目前承载着政府办公功能，各部门位于其附近，包括布鲁宁格孤儿院、市场、旗舰店、市政厅、超市和席勒广场，框定了竞赛区域周边环境的城市形态。

竞赛任务

在城市规划方面，这些"豆子区"和城市中心地带的历史文化遗址亟待翻新。规划区域包括政府办公区、水疗中心、会议酒店以及独立的小型零售商店、咖啡店和餐厅等。规划区域中最引人注目的是堪称纪念物的"银色酒店"，1936—1945年间盖世太保联合中心设立于此。竞赛伊始，为配合"银色酒店"的存在，计划对其周边地块进行一番"脱胎换骨"式的大规模翻新，但随着竞赛的进行，"保留原样"的呼声愈发增强，最终，为保存"银色酒店"而缩减了项目的整体规模。如今，这里又有了一个新名字"多罗逊区"，由 E. Breuninger 有限公司旗下子公司——EKZ Grundstücksverwaltung 有限公司负责其建造事宜。

卡尔广场邻近地区，斯图加特

获奖者

1 一等奖 Behnisch Architekten, Stuttgart (D)
2 二等奖 Kleihues + Kleihues Architekten, Berlin (D)
3 三等奖 Sauerbruch Hutton Architekten, Berlin (D)

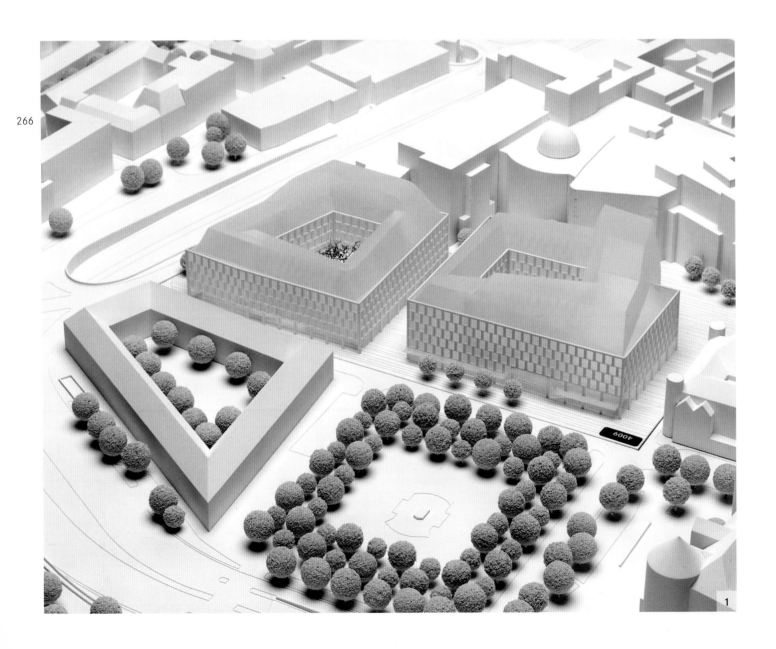

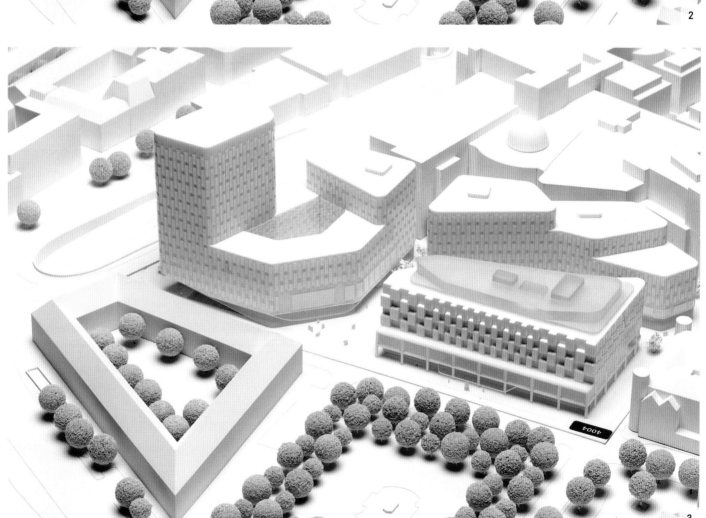

其他参与者

第二轮

4 Blocher Blocher Partners, Stuttgart (D)
5 UNStudio, Amsterdam (NL)
6 Richard Meier & Partner, New York (USA)
7 Rafael Viñoly Architects, New York (USA), with Eller + Eller, Düsseldorf (D)

第一轮

8 gmp Generalplanungsgesellschaft, Frankfurt am Main (D)
9 Henning Larsen Architects, Copenhagen (DK)
10 Bothe Richter Teherani Architekten, Hamburg (D)
11 Foster and Partners and Werner Sobek, London (GB)

竞赛一等奖　获奖事务所

Behnisch Architekten

Stuttgart (D)

方案设计者： Martin Haas
参与者： Theresa Kessler, Stephan Zemmrich, Michael Innerarity, Matthias Neuendorf
顾问： Knippers Helbig, Stuttgart (D); Transsolar Energietechnik, Stuttgart (D); Bartenbach Lichtlabor, Innsbruck (A)

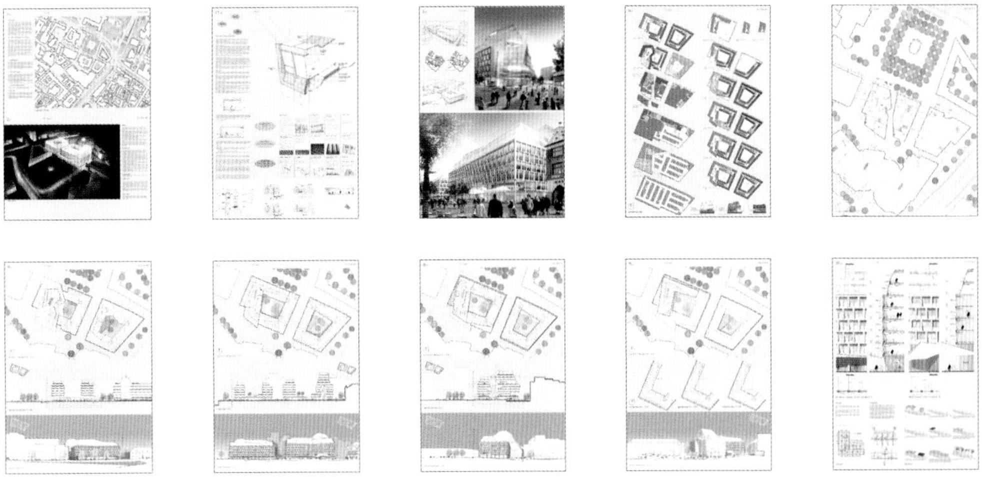

Grundriss Erdgeschoss 1:250

卡尔广场邻近地区，斯图加特

Behnisch Architekten

10/11 Quartier am Karlsplatz — Stuttgart

000646

Ministerium - Fassade / Schnitt / Grundriss 1:50

Hotel - Fassade / Schnitt / Grundriss 1:50

Fassade

Das Licht- und Schattenspiel räumlicher Fassaden bereichert ähnlich den ornamentierten, historischen Fassaden der angrenzenden Gebäude den Straßenraum.

In Materialität und Großzügigkeit orientieren sich die Fassaden aber auch an den schlichten und klaren Bauten der neuen Stadt, des Wiederaufbaus, dem Rathaus und des Marktplatzes.
Überhänge, Vor- und Rücksprünge, schaffen unterschiedliche räumliche Situationen.
Der Straßenraum wird durch die klaren Volumen der Gebäudeskulptur gestärkt, die offenen Erdgeschossfassaden beleben mit ihren Geschäften das neue Quartier am Karlsplatz.

Konstruktion

Das Haupttragwerk ist als Stahlbetonskelettbau mit aussteifenden Kernen konzipiert. Der Skelettbau ermöglicht eine maximale Flexibilität im Ausbau, auch bei zukünftigen Umbauten. Die Deckenspannweiten von i. d. R. 6 - 8 m lassen sich unter Berücksichtigung der geplanten Integration der Haustechnik innerhalb der Deckenplatten mit Flächendeckstärken von 25 bis 30 cm (je nach Spannweite) sehr wirtschaftlich realisieren. Das Untergeschoss wird auf Grund des anstehenden drückenden Wassers mindestens bis zum Bemessungswasserstand als wasserundurchlässige Wanne ausgeführt. Die zurück gesetzten Obergeschosse sind für Gebäudelasten als vorgefertigte Stahlbetonkonstruktion vorgesehen. Durch den Versprung der Stützenachsen im Übergangsbereich sind Abfangungskonstruktionen erforderlich.
Die Gründung erfolgt entsprechend der Empfehlung im Baugrundgutachten als Tiefgründung, die bis auf die Lettenkeuperschicht geführt wird. Auf der Tiefgründung wird zur Sohlabdichtung zunächst eine Unterwasserbetonplatte aus Schwerbeton und zur Herstellung der wasserdichten Baugrubenumschließung sind rückverankerte, überschnittene Bohrpfahlwände vorgesehen.

Darstellung Regelgrundriss 1:200

Prinzipschnitt 1-1 1:50

Tragwerk

Bauablauf

1. Bestand
2. Abbruch der oberirdischen Teile der Bestandsgebäude, Herstellen der seitlichen Baugrubenumschließung (überschnittene Bohrpfahlwand)
3. Abbruch der Bestandsuntergeschosse, Aushub links bis bestehender Kanal, bis auf eine Arbeitsebene oberhalb des Bemessungswasserstandes, parallele Rückverankerung des Baugrubenverbaus, rückverankerte Spundwand zum bestehenden Kanal herstellen
4. Linke Baugrube: Herstellen der Tiefgründung (Bohrpfähle, Rammpfähle, HDI-Säulen, ...) von der Arbeitsebene aus
5. Linke Baugrube: Teilflächenaushub (<500 m²) auf Gründungsniveau, Abschnittweise Herstellen der Unterwasserbetonsohle, vervollständigen der Rückverankerungen des Verbaus
6. Vervollständigen der Sohlplatte links des bestehenden Kanals, Auspumpen der linken Baugrube, Beginn sukzessiver Aushub der rechten Baugrube auf Arbeitsebenenniveau oberhalb des Bemessungswasserstands und freilegen des bestehenden Kanals, Herstellen der Tiefgründung der rechten Baugrube
7. Herstellen des 3. und 2. Untergeschosses inkl. neuem Kanal (WU-Betonwände, Abdichtung, Gefälleboden) auf der linken Seite, Verlegen des Nesenbachs
8. Rückbau des alten Kanals, Vervollständigen des Aushubs rechts, Teilflächenaushub (<500 m²) auf Gründungsniveau, Abschnittsweise Herstellen der Unterwasserbetonsohle, im Bereich der Schnittstelle zwischen linker und rechter Baugrube sukzessiver Aushub und Betoniervorgang direkt anschließend, vervollständigen der Rückverankerungen des Verbaus, Auspumpen der gesamten Baugrube
9. Vervollständigen der restlichen Rohbaus als Stahlbetonskelettbau mit aussteifenden Kernen

Stuttgart

000646

Grundriss 4.Obergeschoss 1:250
+15.00m

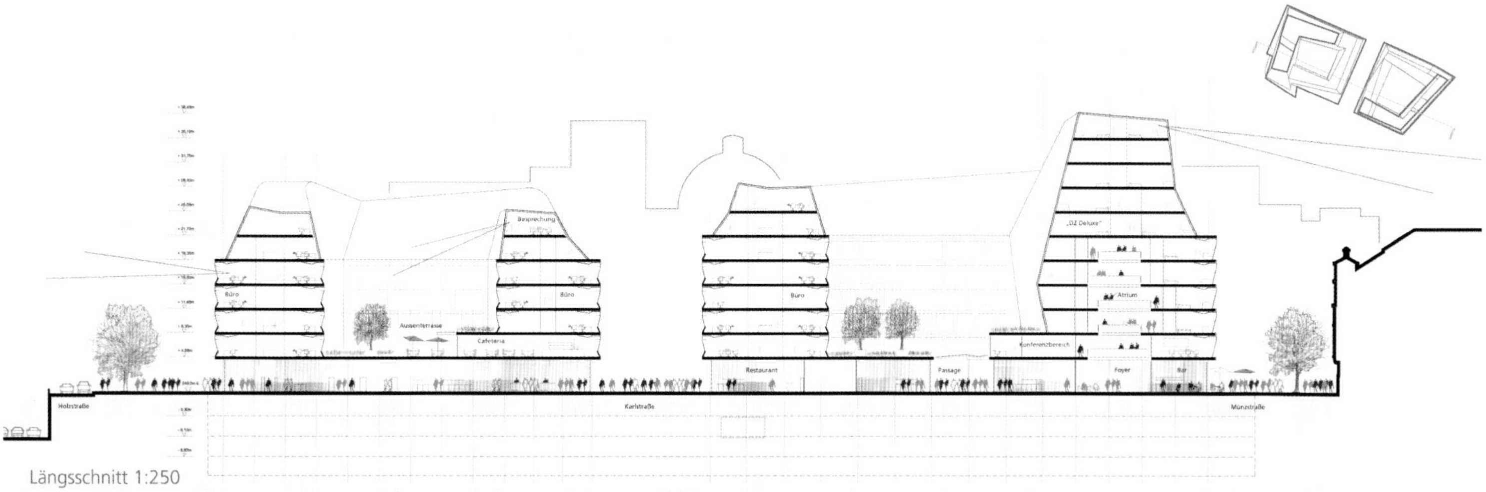

Längsschnitt 1:250

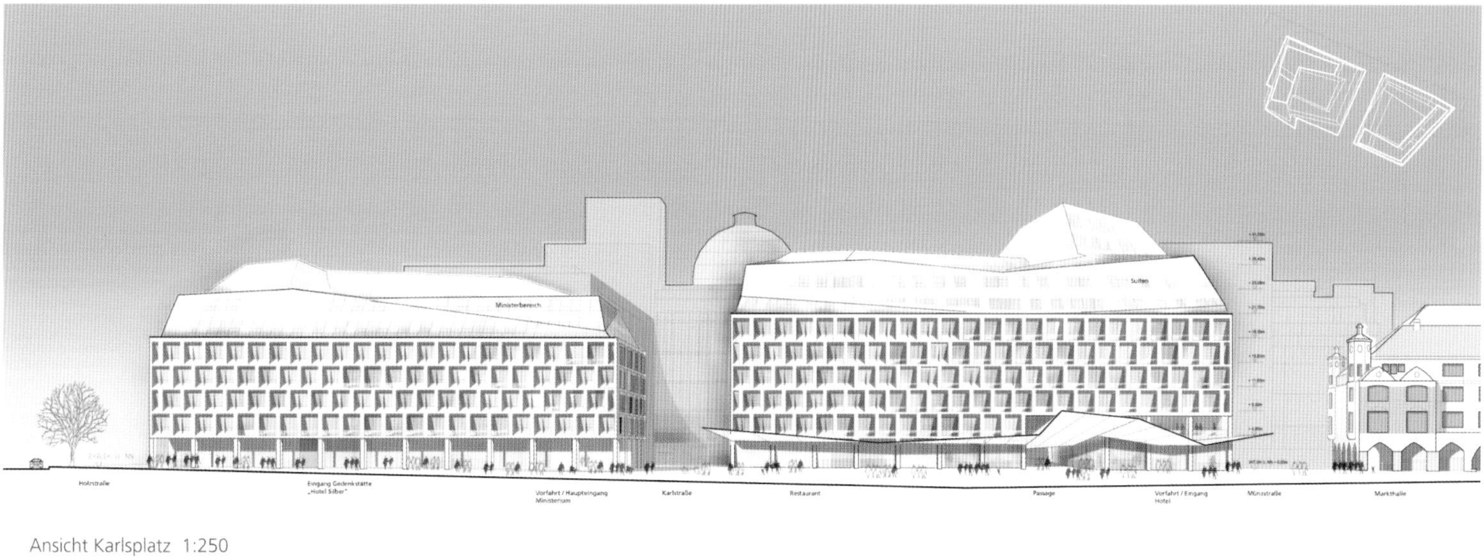

Ansicht Karlsplatz 1:250

Behnisch Architekten

卡尔广场邻近地区，斯图加特

竞赛二等奖 获奖事务所

Kleihues + Kleihues Architekten

Berlin (D)

方案设计者： Jan Kleihues Employees G.tz Kern, Robin Foster, Philipp Zora, Julius St.rmer, Anna Liesicke, Soo-Jin Rim, Philip Schreiber, Daniela Hart, Yasser Shretah, Julia M.ckler
顾问： HL-Technik Engineering Partner, Munich (D), Prof. Klaus Daniels; Boll + Partner, Stuttgart (D); HPP, Munich (D); GRI, Berlin (D); Mosbacher + Roll, Friedrichshafen (D); COMFORT Center Consulting, Düsseldorf (D), Thomas Doerr

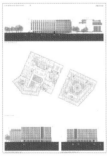
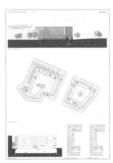
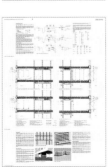
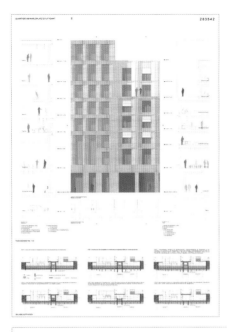
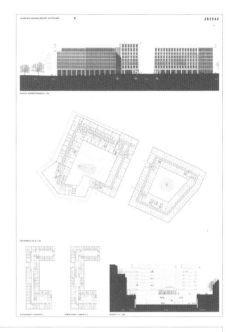
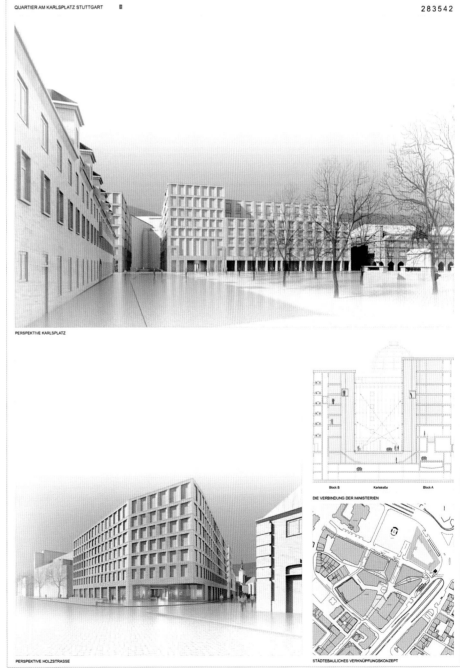

Kleihues + Kleihues Architekten

卡尔广场邻近地区，斯图加特

竞赛三等奖 获奖事务所

Sauerbruch Hutton Architekten

Berlin (D)

方案设计者： Matthias Sauerbruch
参与者： Christian Toechterle-Knuth, Tom Geister, Tarek Ibrahim, Lars Stierwald, Anna Czigler, J.rg Albeke, Claus Marcquardt, Jonas Luther
顾问： Reuter & Rührgartner, Rosbach (D); WSI Frankfurt, Frankfurt am Main (D); HHP Berlin, Berlin (D)

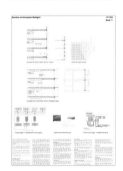
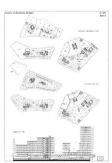

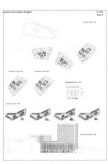
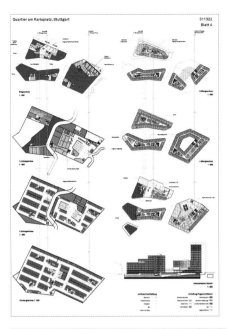
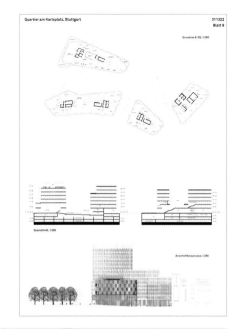
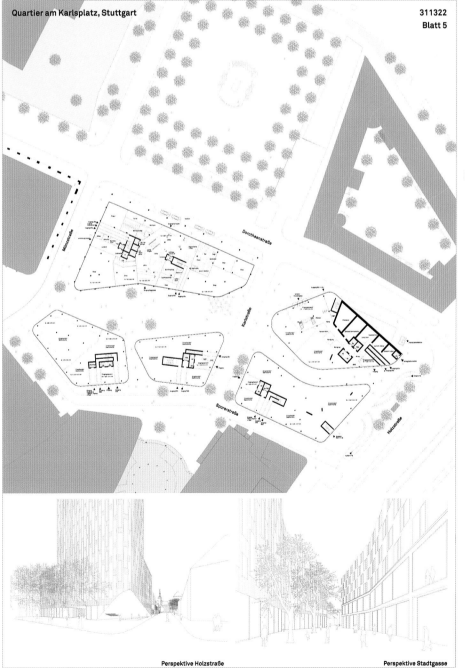

第二轮 获奖事务所

UNStudio

Amsterdam (NL)

方案设计者：Ben van Berkel
参与者：Arjan Dingsté, Marc Herschel, Marianthi Tatari, Marc Hoppermann, Sander Versluis, J.rg Lonkewitz, Kristoph Nowak, Leo Habsburg, Thijst van Zelts
顾问：Ove Arup, Berlin (D); Schlaich Bergermann und Partner, Stuttgart (D)

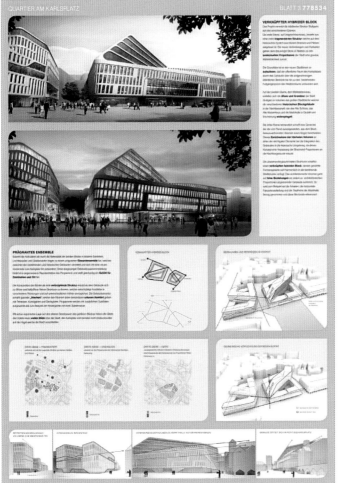
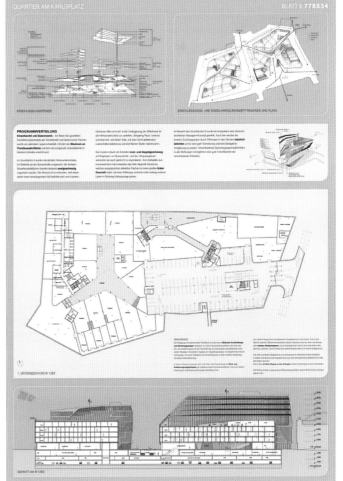

第二轮 获奖事务所

Richard Meier & Partner

New York (USA)

方案设计者： Richard Meier, Bernhard Karpf
参与者： Stefan Scheiber-Loeis, Ringo Offermann, Anne Struewing, Gary Hee, Gil Even-Tsur, Gabriel McKinney
顾问： Arup, New York (USA); Drescher & Partner, Herbolzheim (D)

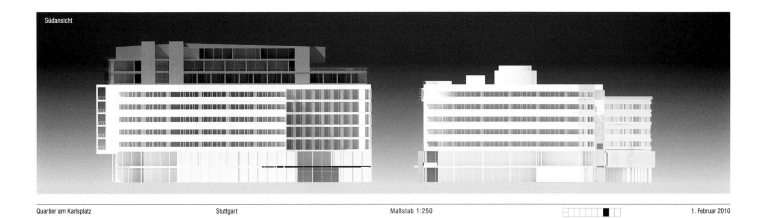
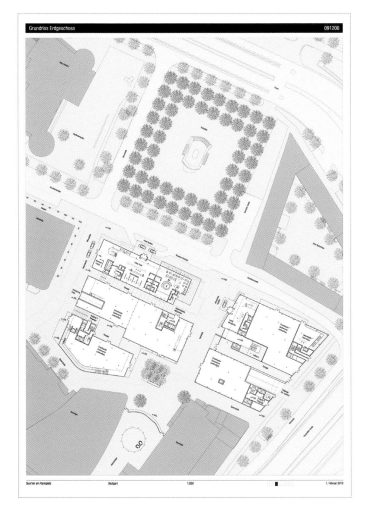
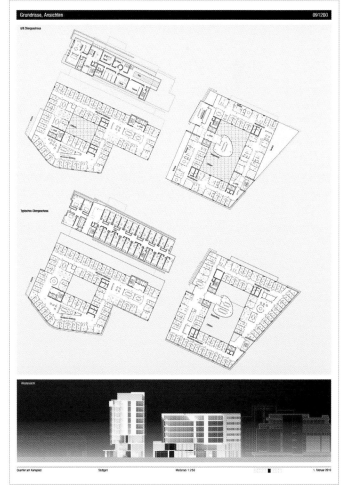

第二轮 获奖事务所

Rafael Viñoly Architects

New York (USA); with Eller + Eller Architekten, Düsseldorf (D)

方案设计者: Jay Bargmann, Erasmus Eller
参与者: Daniel Gartner, Gayley Lambur, Shinichi Horiuchi, Jiwoo Park, Devin Norton, Kuchao Tseng, Satoshi Toyoda
顾问: Mitsu Edwards (F); Mathias Kutterer (D)

- C.1 Einzelhandel
- C.1.1 Einzelhandel
- C.1.2 Gastronomie
- B.5 Hotel Management/Verwaltung
- B.5.1 Buros
- B.5.2 Nebenräume
- B.4.2 Hotel Bar and Lobby

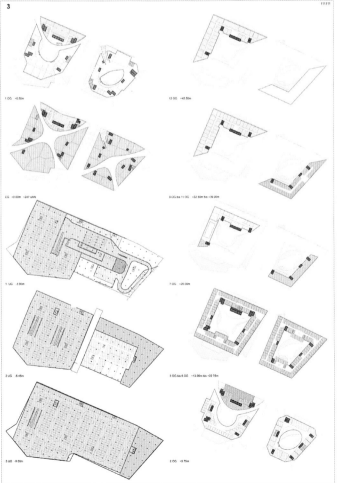

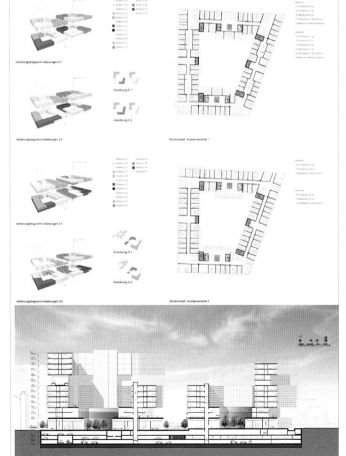

Rafael Viñoly Architects

卡尔广场邻近地区，斯图加特

拉尔日护中心

德国 拉尔

2014

概况

业主：拉尔市
项目规模：基地面积 1500 平方米
类型：开放型设计竞赛（开放申请程序）
参赛者：11 组
竞赛预算：奖金 22 000 欧元

评委

建筑评委：Prof. Jörg Aldinger, Architect, Stuttgart; Prof. Gisela Glass, Architect, Berlin; Prof. Tobias Wulf, Architect, Stuttgart; Prof. Susanne Dürr, Architect, Karlsruhe; Dr. Fred Gresens, Architect, Offenburg; Burkhard Wegener, Landscape Architect, Cologne; Tilman Petters, City Councillor for Building and Construction, Lahr; Silke Kabisch, Facility Management, Lahr
专家评委：Dr. Wolfgang G. Müller, Mayor of Lahr; Guido Schöneboom, Deputy Mayor of Lahr; Rudolf Dörfler, CDU Parliamentary Group, Lahr; Roland Hirsch, SPD Parliamentary Group, Lahr; Jörg Uffelmann, FDP Parliamentary Group, Lahr; Sven Täubert, Bündnis 90/DIE GRÜNEN Parliamentary Group, Lahr; Annerose Deusch, Freie Wähler Parliamentary Group, Lahr

"竞赛赞助商想要建造一座新的建筑，日护中心会聚了三大区域：设有四组儿童日托设施的日常服务区，高卢-罗马半木房子博物馆，当地居民的会议室。建筑室内空间的功能设定遵循了"协同合作效应"。竞赛框架明确了如下问题：以什么形式的特定关系处理历史遗迹？如何解决高卢-罗马居所重建的功能问题？如何设计并打造一个和谐社区？"

——引自《竞赛摘要》

巴登-符腾堡州 2018 园艺展
巴登-符腾堡州 2018 园艺展于德国拉尔市主办。拉尔距离德国弗莱堡大学以北 50 千米。园艺展是该地区城市发展的重要标志。为了配合园艺展的顺利举行，拉尔城市规划提出了几项任务：建造大型公园、日护中心（配备完善的基础设施）、运动场（包含体育馆和多功能厅）。此外，在城市主入口约 47.5 公顷的区域内对以上设施进行升级改造。

大型公园和古罗马 Via Ceramica 遗址
园艺展的规划理念来自德国科隆 L94 俱乐部景观事务所，规划区域包括三个子区域，彼此以一座座桥梁串联在一起。子区域之一是大型公园，可供人们在此社交、健身和娱乐休闲。幼儿园坐落在公园西北部，这里也是古罗马 Via Ceramica 遗址所在地，它是园艺展的一个组成部分。为了保存古罗马 Via Ceramica 遗址，高卢-罗马半木房子博物馆正在重建，植物和其他景观元素均沿着古罗马道路布局的居住区蔓延开来。

日护中心
日护中心的面积大约 1250 平方米，包括三大区域：设有四组儿童日托设施的日常服务区，高卢-罗马半木房子博物馆，当地居民的会议室。三大区域通过共享空间彼此串联（门厅、餐厅和厨房）。日护中心的整体规划在园艺展举行之前完成。在园艺展举行期间，日托中心作为展览的一部分向公众开放；在园艺展结束之后，其作为一个功能多元的公共服务机构，为儿童及其他市民提供服务，是社会公益扶助体系的一部分。

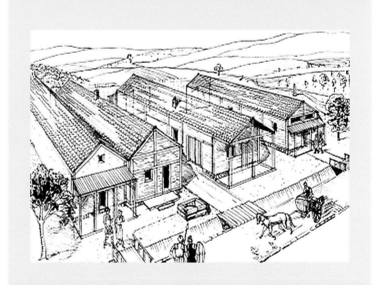

参赛者

获奖者

1 一等奖 (se)arch - Freie Architekten, Stuttgart (D)
2 二等奖 harris + kurrle architekten, Stuttgart (D)
3 三等奖 L/A Liebel/Architekten, Aalen (D)
4 决赛者 Walter Huber Architekten, Stuttgart (D)

第二轮

5 stocker dewes architekten, Freiburg (D)
6 Harter + Kanzler Freie Architekten, Freiburg (D)
7 Ibgo architektur, Munich (D)
8 Lisa Kimling, Freiburg (D)
9 P.I.A – Architekten, Karlsruhe (D)
10 Franke Seiffert, Stuttgart (D)

第一轮

11 Forsthuber & Martinek Architekten, Salzburg (A)

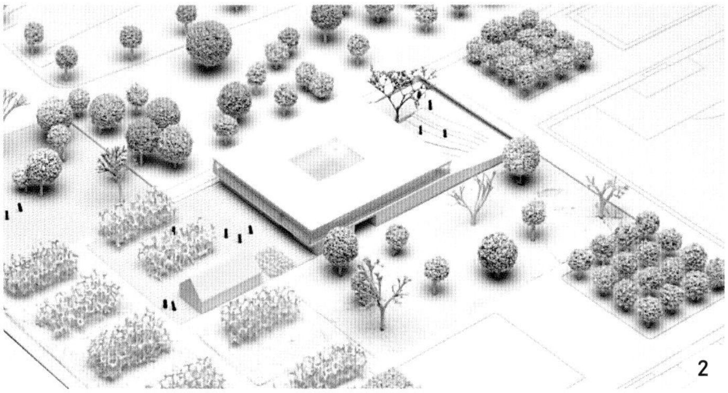

2 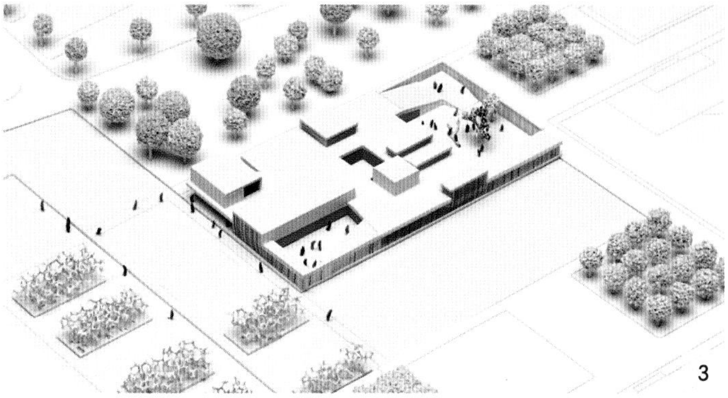 3

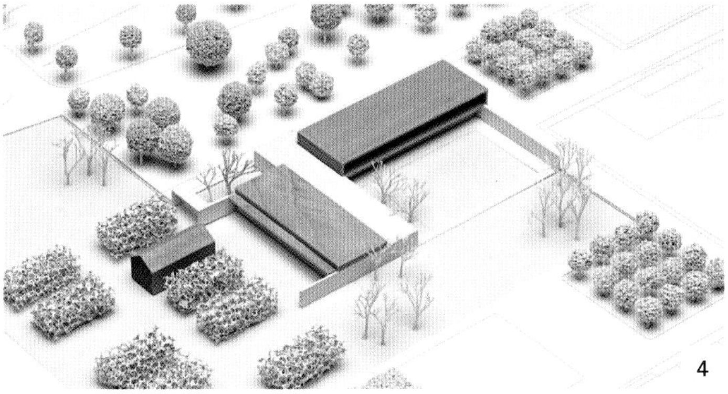

4 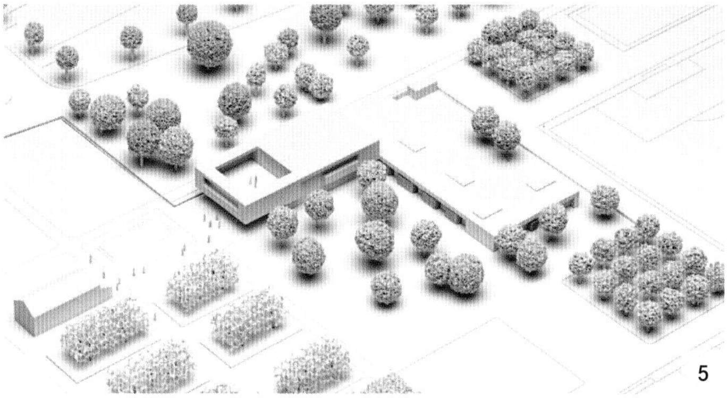 5

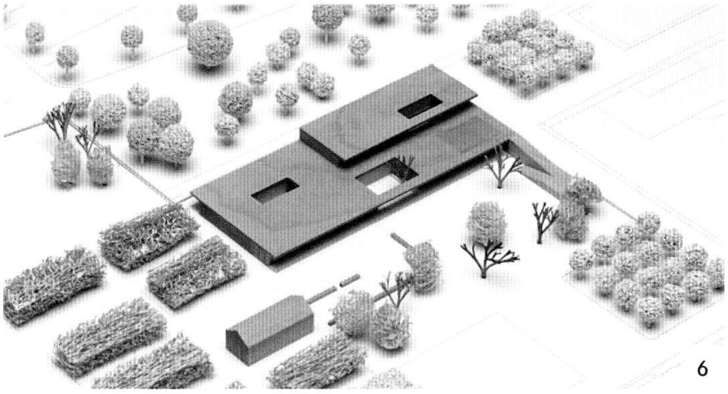

6 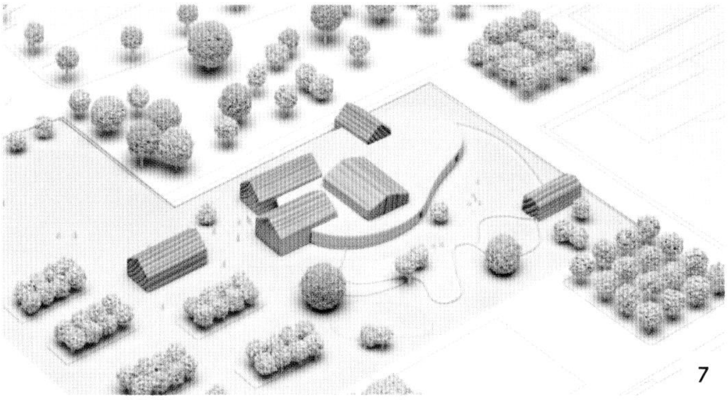 7

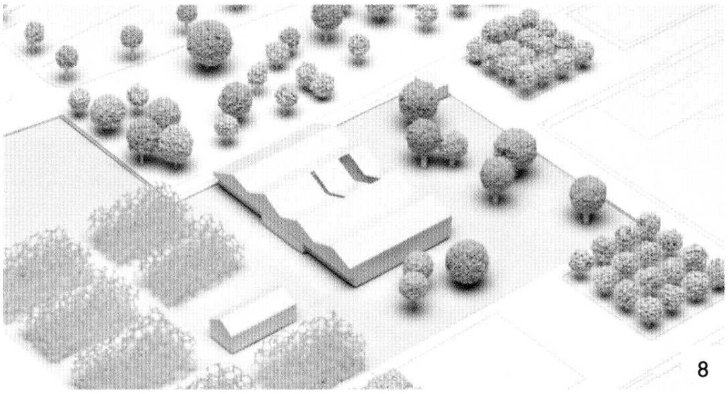

8 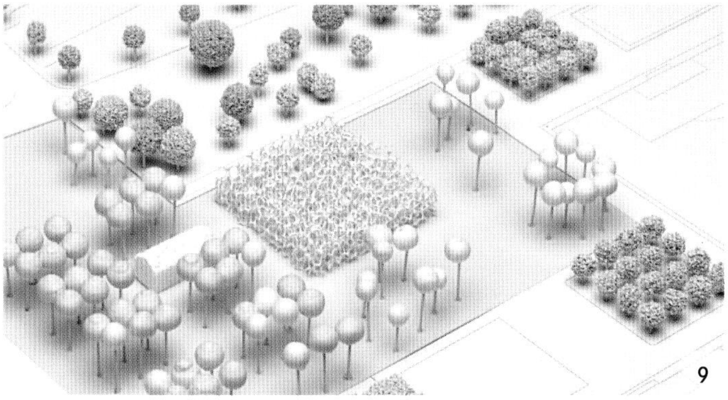 9

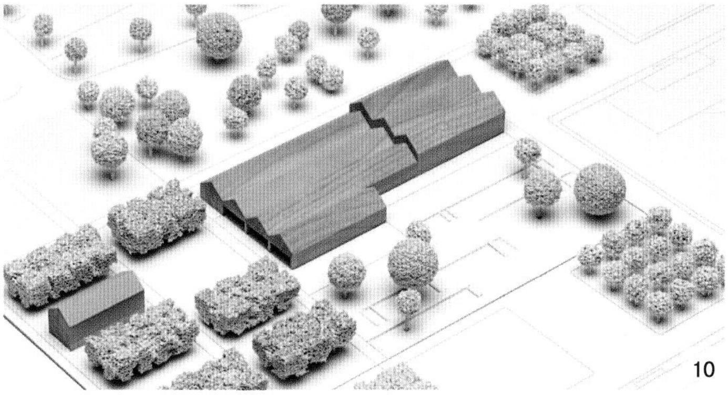

10 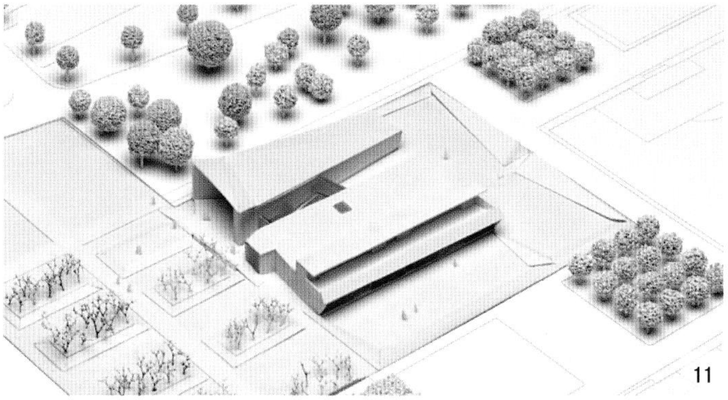 11

获奖者 获奖事务所

(se)arch – Freie Architekten

Stuttgart (D)

方案设计者：Stefanie Eberding, Stephan Eberding
参与者：Sinke Ansis, Stephen O'Brien
顾问：merz kley partner, Dornbirn (A)

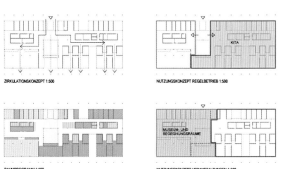
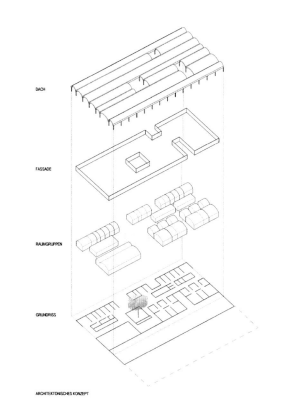
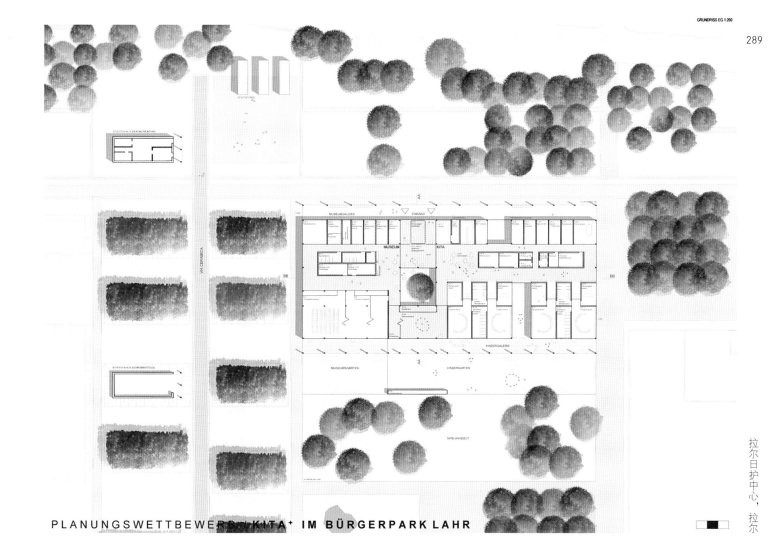

PLANUNGSWETTBEWERB KITA+ IM BÜRGERPARK LAHR

诺伊尔市场（新罗斯托克北部市场）

德国 罗斯托克

2013

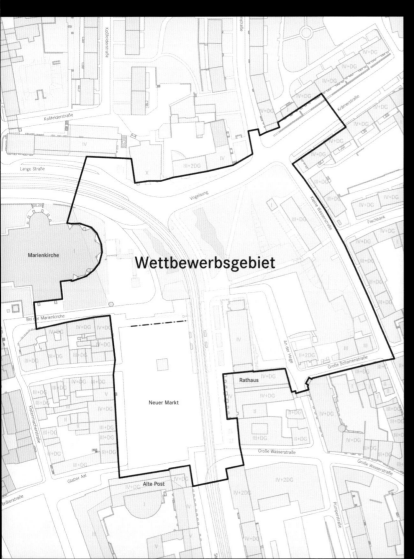

概况

业主：罗斯托克市政府及城市发展和住房管理中心

项目规模：基地面积约 3.75 公顷

类型：双阶段设计竞赛

参赛者：第一阶段 52 组，第二阶段 12 组

竞赛预算：80 000 欧元（第二阶段奖金 70 000 欧元，荣誉奖 10 000 欧元）

评委

建筑评委：Maik Buttler, Architect, Rostock; Prof. Andreas Fritzen, Architect and City Planner, Cologne/Bochum; Prof. Beate Niemann, Architect and City Planner, Düsseldorf/Wismar; Klaus-H. Petersen, Architect, Lübeck; Prof. Kirsten Schemel, Architect, Berlin

专家评委：Roland Methling, Mayor of Rostock; Andreas Herzog, Head of Building Committee, Town Council, Rostock; Andreas Engelmann, Committee for Urban Development, City Assembly, Rostock; Ralph Müller, Department of Urban Development and Economics, Rostock

"诺伊尔市场是德国罗斯托克重要的大型公共市场之一，但其原来的结构布局和部分功能不具有未来发展的潜力。诺伊尔市场的重建工程是城市中心功能完善规划的重要组成部分。"

——引自《竞赛摘要》

罗斯托克
罗斯托克是梅克伦堡-前波美拉尼亚州面积最大和人口最多的城市，有 20.5 万居民。城市从瓦尔诺河向两侧延伸约 20 千米，直至波罗的海。河流是罗斯托克城市风貌的重要组成部分。古城中心位于中世纪城市环形墙的南面。Kröpeliner 城郊住宅区、罗斯托克中央车站和 Steintor-Vorstadt 别墅区位于城墙之外。大规模的战争及之后的重建导致众多历史文化遗址（包括 20 世纪五六十年代的建筑以及 20 世纪 80 年代的大型预制混凝土复合住宅）呈现分散、混杂、小尺度的布局，它们亟待修复或翻新。

诺伊尔市场
约 3.75 公顷的竞赛场地是罗斯托克城市中心区的一部分。该地区的平面布局包括四个部分：兰格街，围绕北部市场的历史文化遗址，待重建的老城区，北部老城区的衔接区域。项目规划旨在改善城市布局，对城市中缺乏组织秩序的区域进行调整，重建并美化顺应规划动线的诺伊尔市场以及市政厅与老城区北部及东部之间的过渡地带。项目规划应该遵循城市历史脉络并顺应二战后城市和建筑的发展潮流。

竞赛任务
重建后的诺伊尔市场集零售商店、公寓群、公共娱乐休闲区于一体，重新成为罗斯托克市的"心脏"。竞赛结果将发展成项目总体规划并为该区域未来的城市和建筑规划策略奠定坚实的基础。

诺伊尔市场（新罗斯托克北部市场），罗斯托克

第一阶段参赛者

第二阶段入围者

1 Schlosser | Schlosser, Berlin (D)
2 Markus Fiegl Architekt with Marek Jahnke Landschaftsarchitekt, Berlin (D)
3 DE+ Architekten, Berlin (D)
4 SMAQ - architecture urbanism research, Berlin (D)
5 Steiner Weißenberger Architekten, Berlin (D)
6 Heine Mildner Architekten, Dresden (D)
7 STOY - Architekten, Neumünster (D)
8 Hübotter + Stürken Architektengemeinschaft with Lohaus + Carl Landschaftsarchitekten,
9 Gnadler.Meyn.Woitassek Architekten Innenarchitekten, Stralsund (D)
10 mhb Planungs- und Ingenieurgesellschaft, Rostock (D)
11 Lips + Teichert Architekten, Freiburg-March (D)
12 matrix architektur, Rostock (D), with Kirk + Specht Landschaftsarchitekten, Berlin (D)

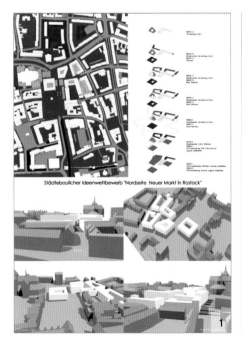
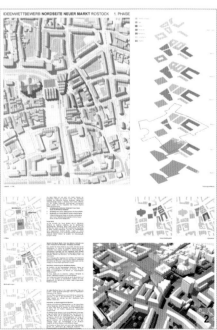
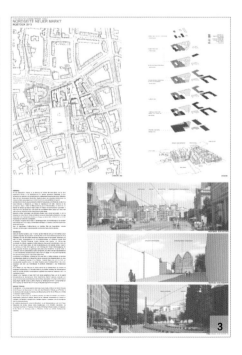

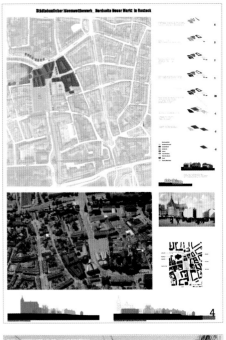

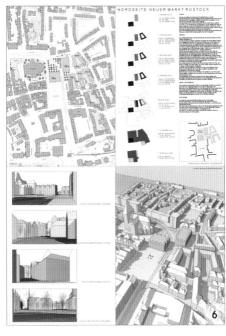
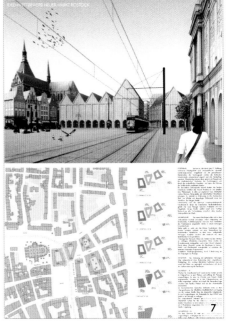
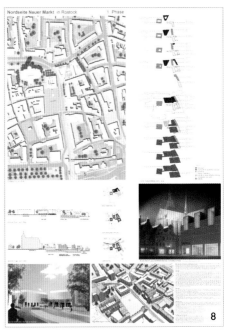

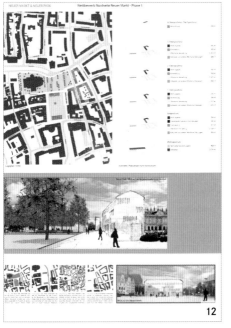

第一阶段其他参与者

13 Zvi Hecker Architekt, Berlin (D) **14** Freie Architektin Jeong, Stuttgart (D) **15** Architekturbüro Borries + Partner, Rostock (D) **16** Uwe Becker, Berlin (D) **17** Müller-Born-Architekten, Kassel (D) **18** matzke | architekten, Berlin (D) **19** INROS LACKNER, Rostock (D) **20** Art de Lux, Munich (D) **21** Alexander Paul Architekt, Glienicke (D) **22** Schultes Frank Architekten, Berlin (D) **23** Susanne Dieckmann Architektin, Weimar (D) **24** Deubzer König Architekten, Berlin (D) **25** Geske-Wenzel Architekten, Berlin (D) **26** MRSCHMIDT Architekten, Berlin (D) **27** Gauselmann Soll Architekten, Dortmund (D) **28** Büro Luchterhandt Stadtplanung, Hamburg (D) **29** Czerner Göttsch Architekten, Hamburg (D) **30** Roland Unterbusch Architekt, Rostock (D), with Gesa Königstein Landschaftsarchitektin, Berlin (D) **31** Gasparini Christian Architekt, Reggio Emilia (I) **32** Frank Görge Architekt, Hamburg (D) **33** argeplan, Hanover (D) **34** Franck – Von Reusner, Lübeck (D) **35** r10r10 Architects, Stuttgart (D) **36** Volkmar Nickol, Berlin (D) **37** Wiencke Architekten, Dresden (D) **38** Thomas Möller Architekt, Karlsruhe (D) **39** DE FABRIEK, Nijmegen (NL), with Thus Ton Städtebauer & Landschaftsarchitekt, Ubbergen (NL), and Akkers Joop Architekt, Wijchen (NL) **40** Janin Rabaschus Architektin, Dresden (D) **41** fabriK·B Architekten and Scharf und Wolf, Berlin (D) **42** Locke Lührs Architektinnen, Dresden (D) **43** Köppler Türk Architekten, Berlin (D) **44** Architekturbüro Albert und Planer, Rostock (D) **45** Frank Architekten, Munich (D) **46** J. Deutler/Architekturbüro R. Schacht, Rostock (D) **47** T.A. Wolf Architekten, Munich (D) **48** Josep Sánchez Ferré Architekt, Barcelona (E) **49** Pannett & Locher Architekten, Bern (CH) **50** Nopto Architekt, Herzebrock-Clarholz (D) **51** Kai Lorberg Architekt, Hamburg (D) **52** BBCF Architectes, Paris (F)

第二阶段参赛者

获奖者

1. DE+ Architekten, Berlin (D)
2. Hübotter + Stürken Architektengemeinschaft, Hanover (D)
3. SMAQ – architecture urbanism research, Berlin (D)

荣誉奖

4. Steiner Weißenberger Architekten, Berlin (D)
5. mhb Planungs- und Ingenieurgesellschaft, Rostock (D)

第二轮

6. Gnadler.Meyn.Woitassek Architekten Innenarchitekten, Stralsund (D)
7. matrix architektur, Rostock (D), with Kirk + Specht Landschaftsarchitekten, Berlin (D)
8. Schlosser | Schlosser, Berlin (D)
9. Markus Fiegl Architekt with Marek Jahnke Landschaftsarchitekt, Berlin (D)
10. Lips + Teichert Architekten, Freiburg-March (D)
11. Heine Mildner Architekten, Dresden (D)

第一轮

12. STOY – Architekten, Neumünster (D)

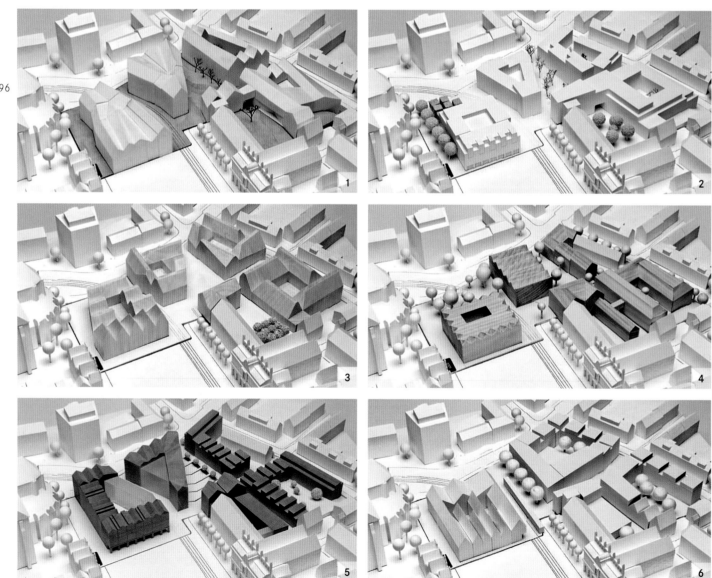

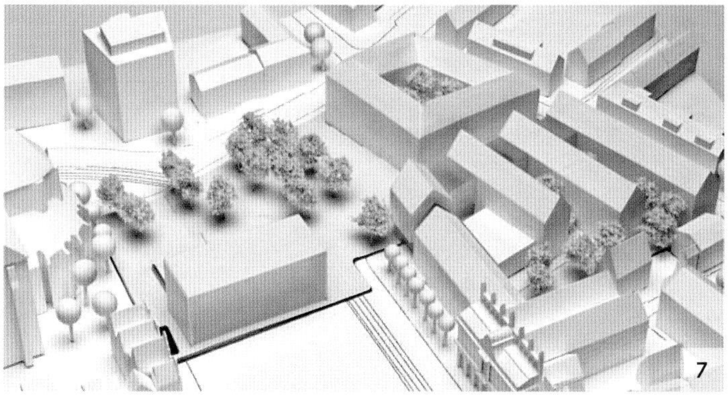

第二阶段参赛者

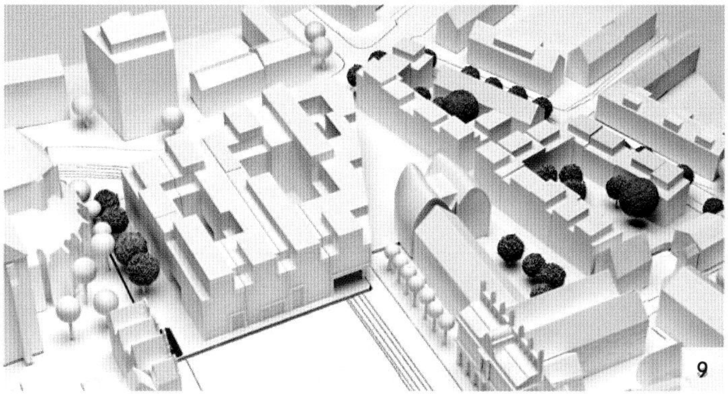

297

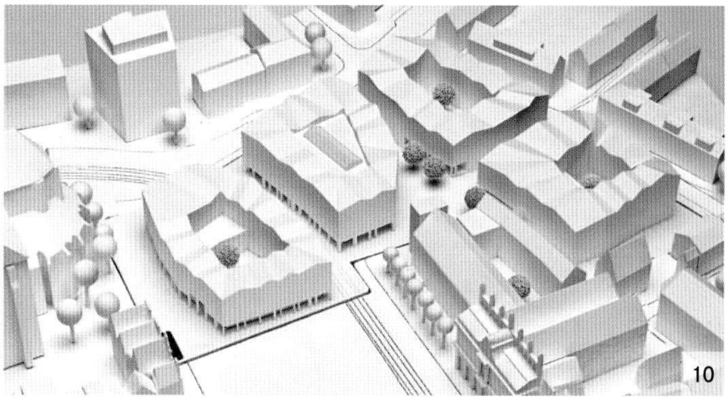

诺伊尔市场（新罗斯托克北部市场），罗斯托克

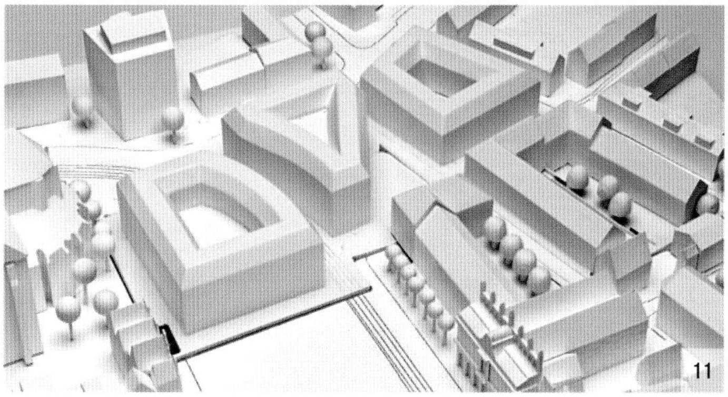

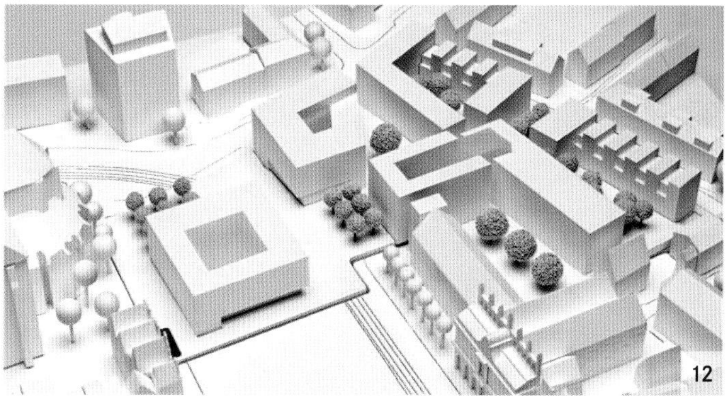

住宅 F

德国 柏林

2009

概况
业主：私人业主
项目规模：基地面积约 657 平方米
类型：邀标合作型设计竞赛
参赛者：6 组
竞赛预算：30 000 欧元（奖金 15 000 欧元，2500 欧元）

评委
建筑评委：Winfried Brenne, Architect, Berlin; Pro Architect, Hanover; Jürgen Mayer H, Architect, Berli
专家评委：Project Clients, Berlin

"我希望房子是一个很大的心形。房子里有一个花园。我午饭吃面条。阿K睡在房顶，我睡在下面。"（业主大女儿的梦想）

"我希望房子是一颗星星，星星的每个角都有一间房子。一个大卧室供所有孩子使用，我们睡在一个房间里。一间五彩缤纷的桑拿浴室，一个五彩缤纷的浴缸，一对兔子。"（业主小女儿的梦想）

——引自《竞赛摘要》

场地
项目位于德国柏林西南部，大约2000平方米。城市框架形成了一个同质化且开放的别墅区以及城乡接合部的小型家庭住宅。在20世纪70—90年代，建筑与基地相邻，基地上的树木需要重点保护。

程序
竞赛内容非常普通，但在以往竞赛中却极其罕见。竞赛对象仅为一个家庭。业主对建筑设计有浓厚的兴趣，他们的好奇心和较少的限制促成了一种经典的竞赛形式，即建筑师以匿名的方式同时作为参赛者和评审者。

竞赛任务
除了约650平方米（包含地下室）的房子，房子周边橘园和花园的规划也非常重要；业主爱好园艺，橘园和花园是其最爱。常规的起居室、餐厅、书房、五个孩子共享的大空间、会客厅等均应被充分考虑。竞赛任务是以规划功能多元的空间为目标，满足业主的空间使用需求；在此基础上，在现代化的建筑环境中融入传统乡村式的自然景观，在乡村风格的景观中融入当代建筑语汇；房子的建造流程既符合日益严格的建筑法律法规的相关规定，又最大限度地利用现有资源；室内外空间连接紧凑，交通动线仿佛是"动态流动"的。总而言之，业主想要的并非一个"高科技房子"，而是一个既漂亮又实用的房子，一切考虑均"以人为本"。

住宅F，柏林

获奖者

1 一等奖 Delugan Meissl Associated Architects, Vienna (A)
2 二等奖 augustin und frank architekten, Berlin (D)
3 三等奖 Staab Architekten, Berlin (D)

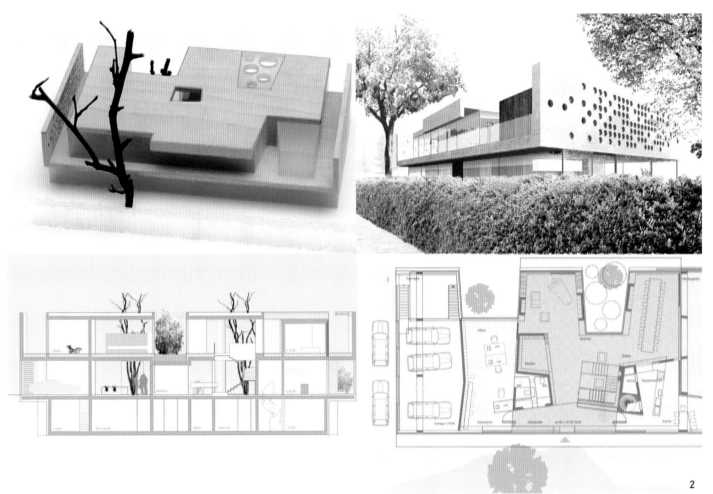

其他参与者

4　决赛者 Graft Architekten, Berlin (D)

第三轮
5　agps architecture, Zurich (CH)/Los Angeles (USA)
6　Dolenc Scheiwiller Architekten, Zurich (CH)

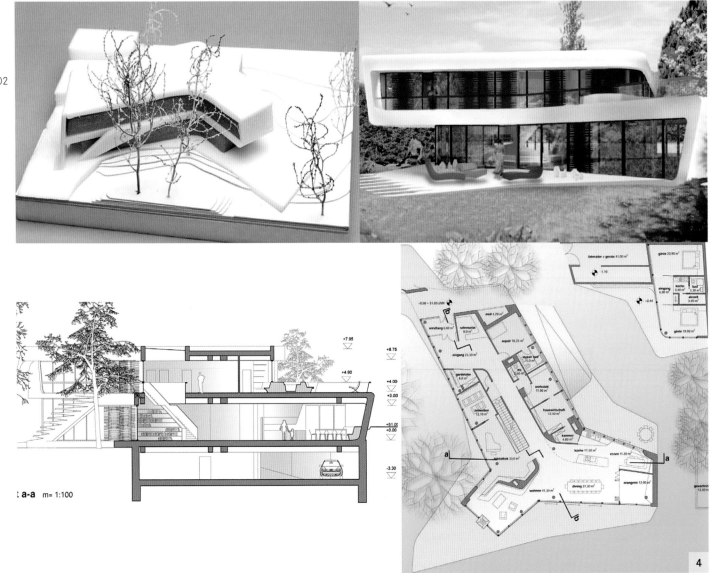

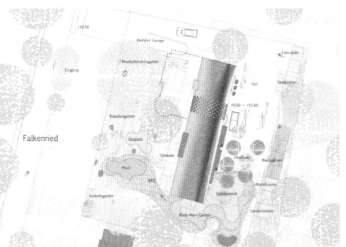

竞赛一等奖　获奖事务所

Delugan Meissl Associated Architects (DMAA)

Vienna (A)

方案设计者：Roman Delugan, Elke Meissl
参与者：Anna Edthofer, Michael Lohmann, Claudia Schiedt, Christian Schrepfer
顾问：Werner Sobek, Stuttgart (D); Rajek + Barosch, Vienna (A)

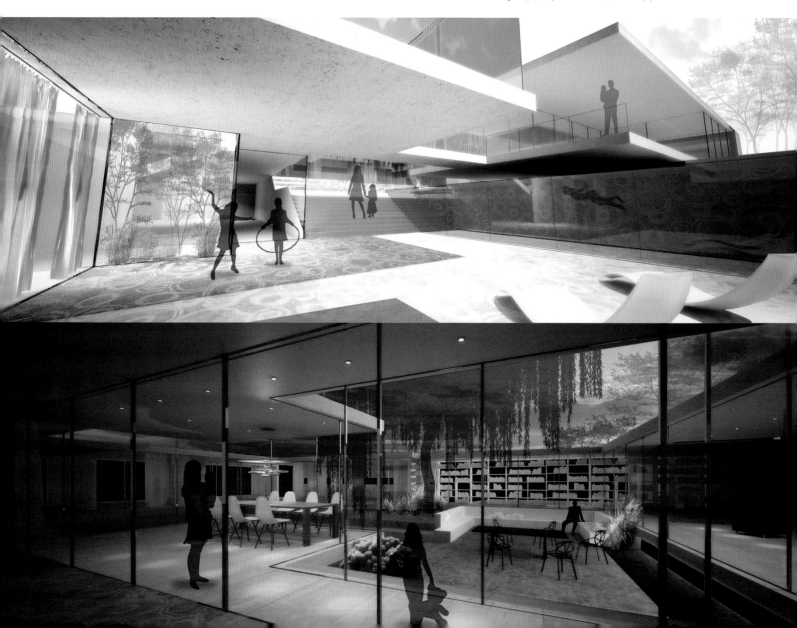

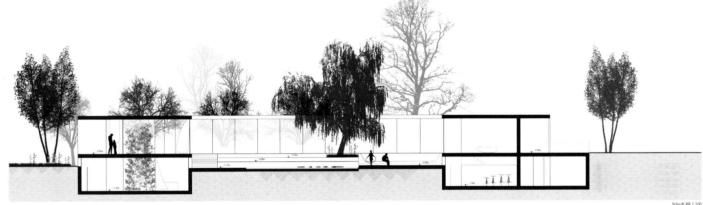
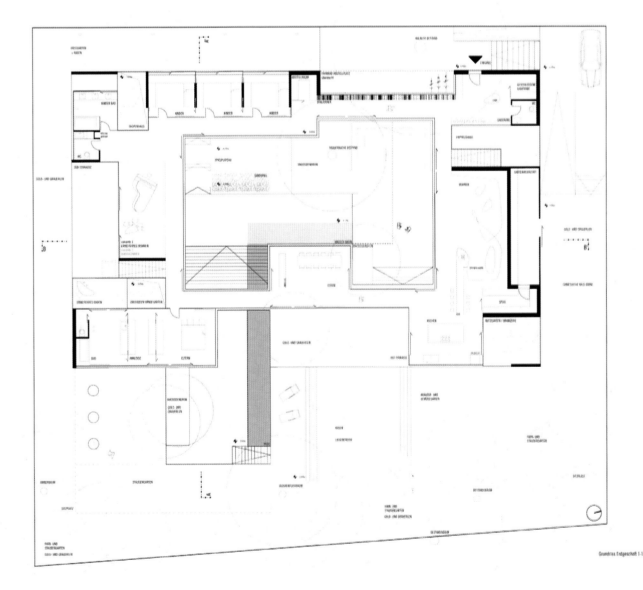

"建筑的艺术"
——勃兰登堡州议会大厦

德国 波茨坦

2011—2012

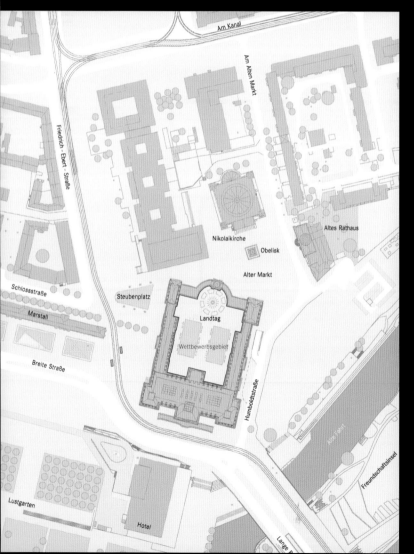

概况

业主：勃兰登堡州议会和财政部
项目规模：院落约 74 m×57 m，基地面积 4200 平方米
类型：双阶段艺术竞赛
参赛者：第一阶段 100 组，第二阶段 11 组
竞赛预算：480 000 欧元

评委

艺术评委：Leonie Baumann, Rector, Berlin Weissensee School of Art; Prof. Michael Braum, Chairman of the Board, Bundesstiftung Baukultur, Potsdam; Daniela Dietsche, Managing Director, Brandenburgischer Verband Bildender Künstlerinnen und Künstler e.V., Potsdam; Prof. Mischa Kuball, Independent Artist and Professor of Holography/Light Art, Academy of Media Arts Cologne; Maria Ossowski, Head of the Department for Culture, Rundfunk Berlin-Brandenburg, Berlin; Sabina Grzimek, independent artist, Berlin; Prof. Dr. Brigitte Rieger-Jähner, Director, Museum Junge Kunst Frankfurt (Oder); Dr. Barbara Steiner, Director, Galerie für Zeitgenössische Kunst, Leipzig

专家评委：Gerrit Große, Vice President of the State Parliament, Brandenburg State Parliament Member (DIE LINKE) and Chairman of the Brandenburg State Art and Facilities Committee; Anja Heinrich, Brandenburg State Parliament Member (CDU), Member of the Brandenburg State Art and Facilities Committee; Marie-Luise von Halem, Brandenburg State Parliament Member (BÜNDNIS 90/DIE GRÜNEN), Member of the Brandenburg State Art and Facilities Committee; Marianne Kliem, Department 4 - State Assets, Salary Law, Property Management and Construction Administration, Staff Unit for the New State Parliament Building, Brandenburg Ministry of Finance; Prof. Peter Kulka, Architekt, Dresden/Cologne; Jens Lipsdorf, Brandenburg State Parliament member (FDP), Member of the Brandenburg State Art and Facilities Committee; Susanne Melior, Brandenburg State Parliament Member (SPD), Member of the Brandenburg State Art and Facilities Committee; Michael Ranft, Head of Administrative Department V, Brandenburg State Parliament

"艺术不需要反映历史,建筑的艺术不必对建筑的内容和目的进行评价。"

——引自《竞赛摘要》

勃兰登堡州议会

2005 年 5 月,德国勃兰登堡州议会通过决议,在 17 世纪城市宫殿的基础上构建一座议会大厦。城市宫殿在二战期间遭到损坏并在之后一个世纪中几乎被完全摧毁。2011 年春季,该项目举行奠基仪式。2013 年底,议会大厦的外立面和外轮廓已经完成;2014 年初,该项目落成并交付使用,议会成员与工作人员陆续搬入。议会大厦由德累斯顿建筑事务所的建筑师 Peter Kulka 设计。规划和建设方案在实施前,关于重建由建筑师 Georg Wenceslaus von Knobelsdorff 专为国王弗雷德里克二世设计的城堡的合理性及其具体事宜经过了数年激烈的争论,这是民主政治制度的一部分。

"建筑的艺术"

这是一场有关艺术的竞赛。艺术作品可能是议会自我形象的讨论,可能是公民政治决策的认知,也可能是公共空间的议会展示。艺术不需要反映历史,建筑的艺术不必对建筑的内容和目的进行评价。然而,竞赛主办方认为,鉴于项目的特殊性质及其所在的位置,对"建筑的艺术"进行研究是非常有必要的。

竞赛任务

在议会大厦的庭院中,一系列艺术作品被构思和布置;它们专为议会大厦而设,确切地反映议会大厦及其场地的文脉关系。艺术作品的设计理念与场地、建设任务、历史文脉、议会机构以及功能职责相结合。此外,纯粹的装饰方法被批判性地审视。从长远考虑,议会大厦作为城市公共建筑,应该不断地提升公共服务功能,从而为城市区域复兴作出贡献。

「建筑的艺术」——勃兰登堡州议会大厦,波茨坦

第二阶段参赛者

获奖者

1　一等奖 Florian Dombois, Cologne (D)
2　二等奖 Annette Paul, Potsdam (D)
3　三等奖 Hester Oerlemans, Berlin (D)

第二轮

4　Heinke Haberland, Düsseldorf (D)
5　Gerald Hofmann, Nuremberg (D)
6　Peter Sandhaus, Berlin (D)
7　Götz Lemberg, Berlin (D)
8　Thomas Eller and Füsun Türetken, Berlin (D)
9　Marc Bausback, Berlin (D)
10　Markus Klink, Stuttgart (D)
11　Christoph Faulhaber, Hamburg (D)

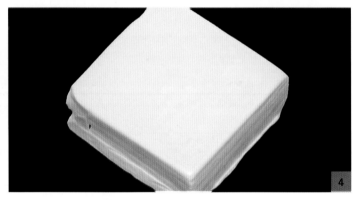

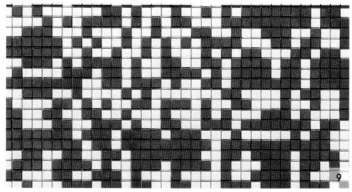

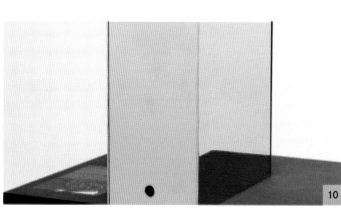

吕讷堡大学校园景观设计

德国 吕讷堡

2010

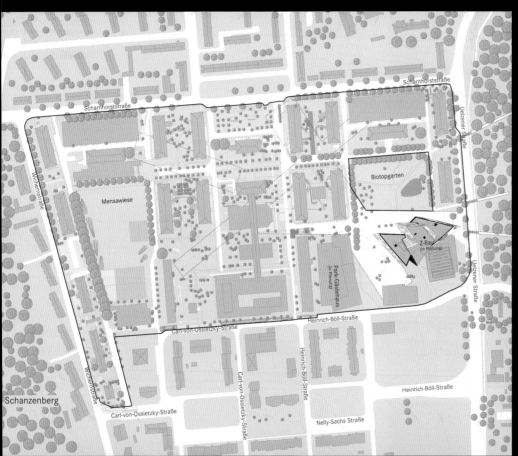

概况
业主：吕讷堡大学基金会
项目规模：基地面积约 15 公顷
类型：开放型设计竞赛（开放申请程序）
参赛者：6 组
竞赛预算：48 000 欧元（奖金 18 000 欧元，每组参赛者的费用 5000 欧元）

评委
建筑评委：Prof. Daniel Libeskind, Architect New York; Tobias Micke, Landscape Architect

"大学校园作为地标和充满活力的文化生活中心,与新建的中心建筑共同彰显大学甚至城市的特色,这是高等学府校园规划的重要愿景及其未来发展的方向所在。"
——引自《竞赛摘要》

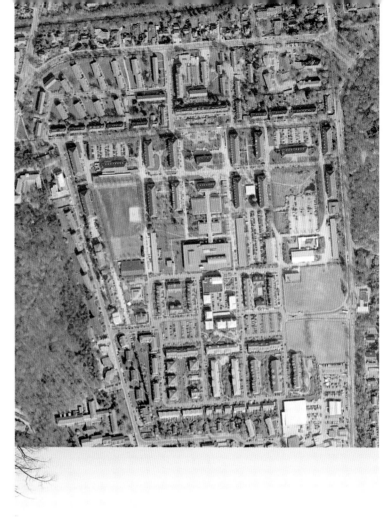

吕讷堡大学
吕讷堡大学是一所示范大学,教学和科研是其两大功能,并设立多项奖、助学金;其多年来的教学和科研成果已成为该区域经济发展的重要推动力。目前,吕讷堡大学有三个不同的校区。中央校区所在地块(约16公顷)曾在20世纪30年代后期作为沙恩霍斯特军营。基地内遍布绿化带和林荫小巷,处于温泉公园和伊尔瑙谷地和之间。现有建筑以轴网严密的砖造兵营建筑为特色。

丹尼尔·里伯斯金校区的校园规划策略
丹尼尔·里伯斯金校区整体校园及中央建筑的规划策略在深入推敲现有地块体系结构的基础上,形成一个"对位布局(counterpoint)"。

竞赛任务
竞赛任务的核心是为约16公顷的中央校区及其户外基础设施进行整体性规划。参赛者所提交的校园规划必须是灵活的、全面的、可操作性强的、具有未来发展潜力的,即践行"稳健发展"的设计理念。校园规划可分几个施工阶段逐步实施。建筑师面临的挑战一方面是充分体现历史建筑网格和林荫小路之间的区别,另一方面是与丹尼尔·里伯斯金校区"对位布局(counterpoint)"的校园规划策略形成呼应,并通过两条轴线将中央校区和丹尼尔·里伯斯金校区串联在一起。户外基础设施应该是现代化的,并方便行人和机动车通行。此外,校园规划应体现吕讷堡大学的特色教学和科研领域——可持续发展、文化、教育,同时符合大学管理和运营标准。

吕讷堡大学校园景观设计,吕讷堡

参赛者

获奖者

1 一等奖 Karres en Brands Landschapsarchitecten,
  Hilversum (NL)
2 二等奖 Weidinger Landschaftsarchitekten, Berlin (D)
3 三等奖 Breimann & Bruun Landschaftsarchitekten,
  Hamburg (D)

第一轮

4 geskes.hack Landschaftsarchitekten,
  Berlin (D)
5 el:ch landschaftsarchitekten,
  Berlin (D)
6 gh3,
  Toronto (CDN)

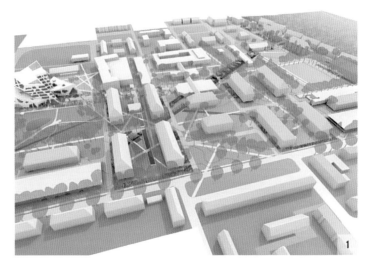

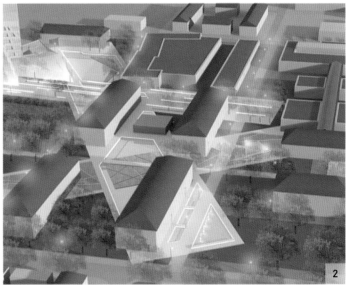

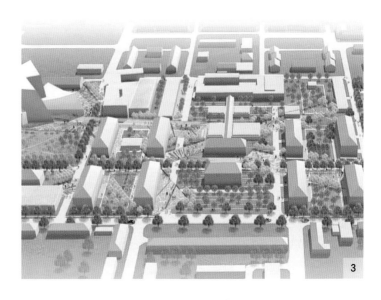

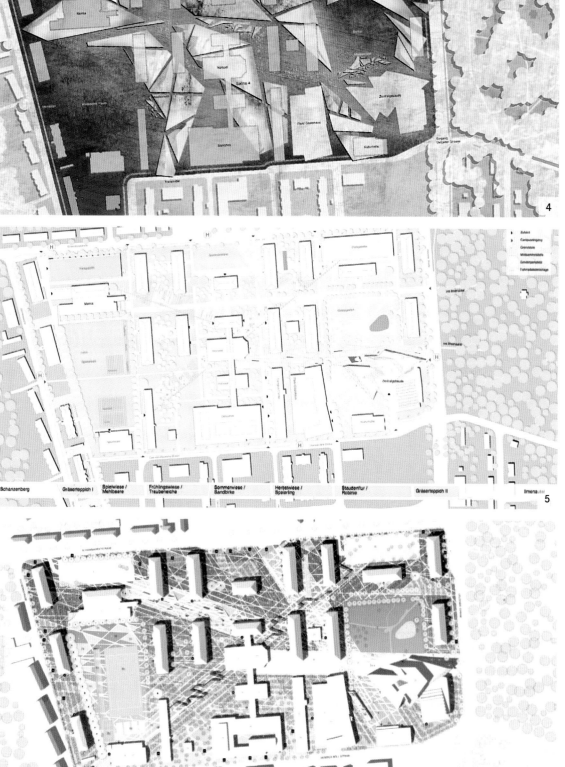

竞赛一等奖　获奖事务所

Karres en Brands Landschapsarchitecten

Hilversum (NL)

方案设计者：Bart Brands, Sylvia Karres
参与者：Milda Jusaite, Davinder Lota, Sascha Seidel, Eddie Silveira, Anna Sobiech

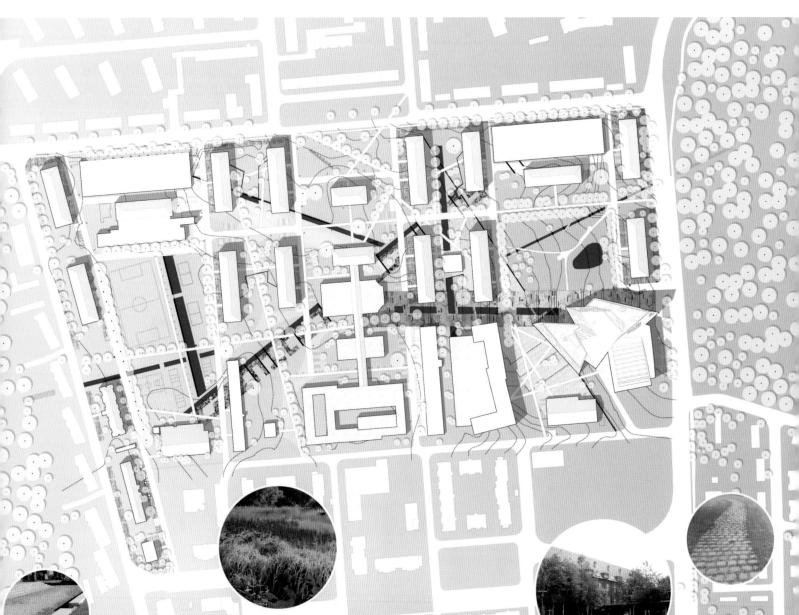

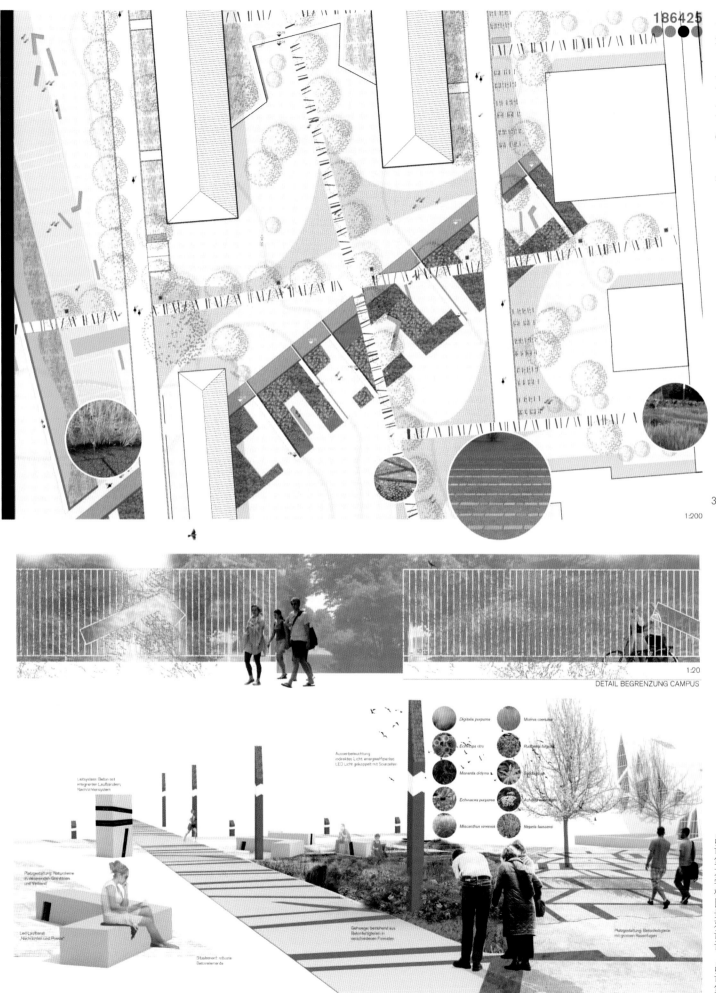

1:200

DETAIL BEGRENZUNG CAMPUS
1:20

DETAILS AUSSTATTUNG, MATERIAL, PFLANZEN UND FARBEN

竞赛二等奖 获奖事务所

Weidinger Landschaftsarchitekten

Berlin (D)

方案设计者：Jürgen Weidinger
参与者：Paul Giencke, Ralf Kammeyer, Guohao Li, Julia Patzak
顾问：Ingenieurbüro Kraft Beratende Ingenieure für Wasserwirtschaft, Berlin (D)

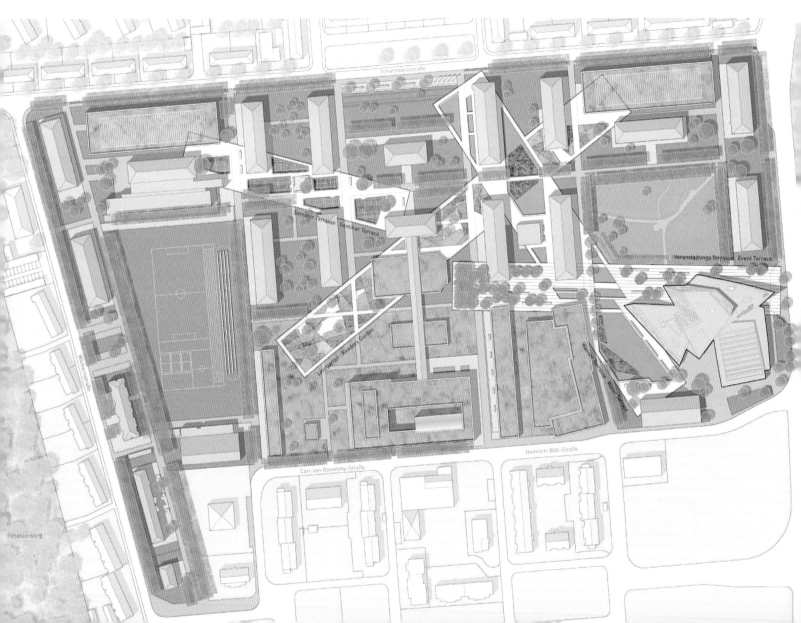

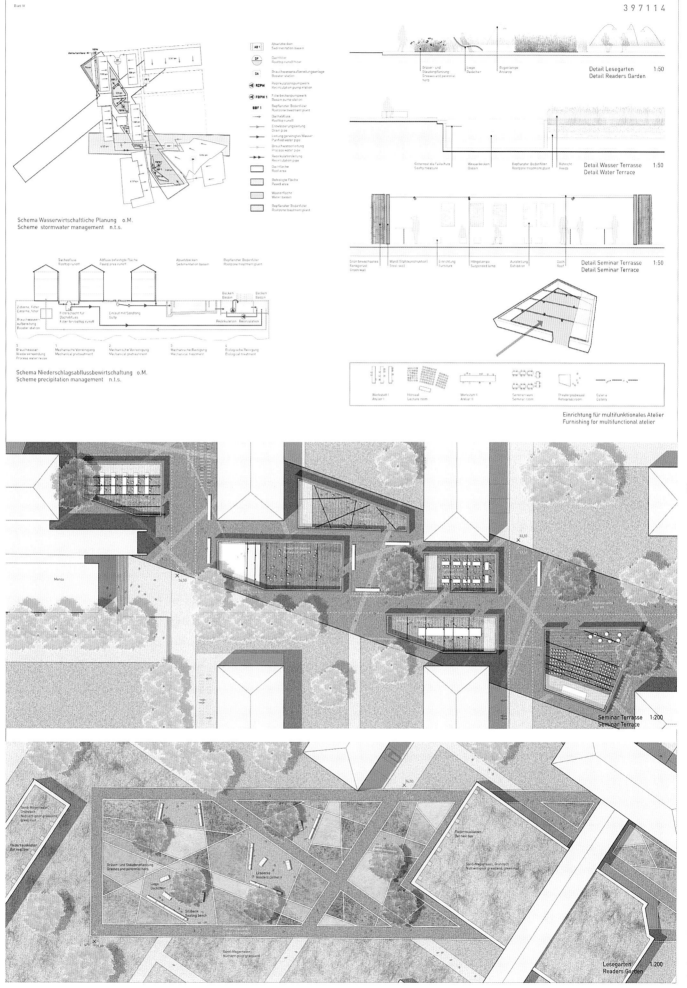

竞赛三等奖 获奖事务所

Breimann & Bruun Landschaftsarchitekten

Hamburg (D)

方案设计者：Judith Haas, Moritz M.llers
自由职业者：Magdalena Cieslicka

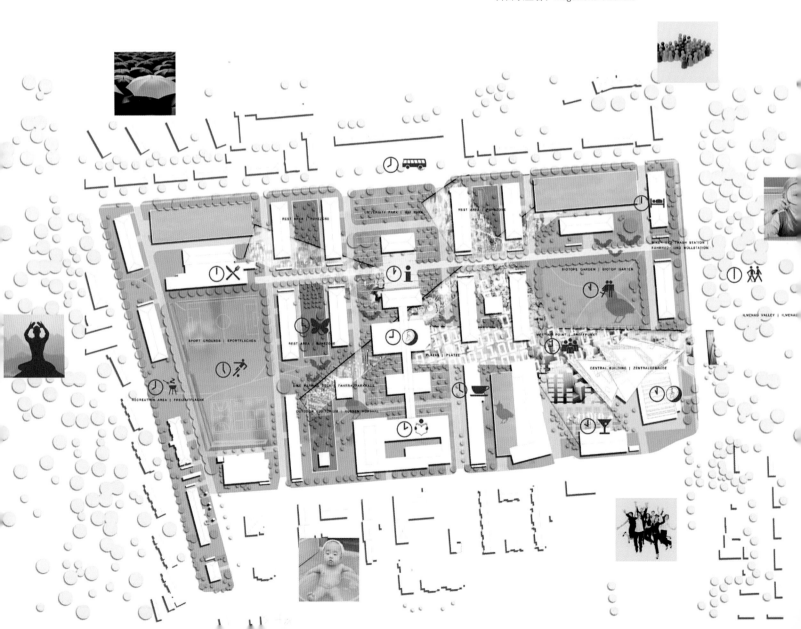

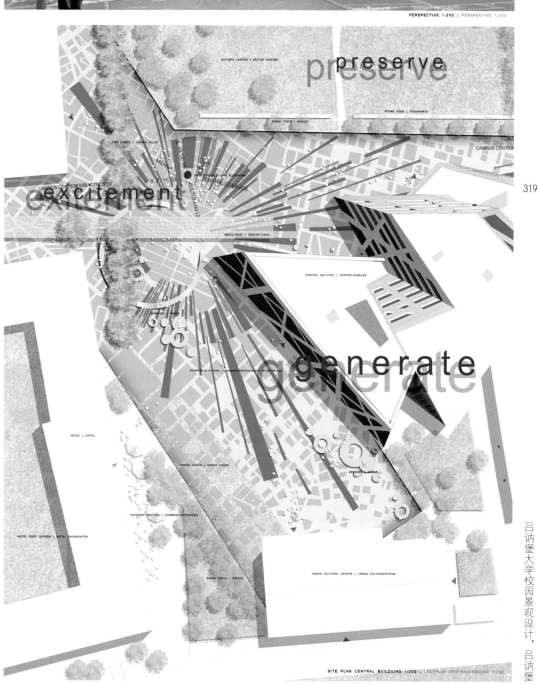

新贝斯科艺术档案馆

德国 贝斯科

2009—2010

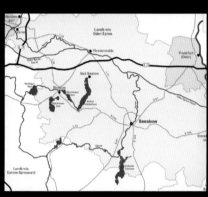

概况
业主：贝斯科市，施普雷区
项目规模：基地面积 3000 平方米
类型：开放型设计竞赛（开放申请程序）
参赛者：15 组
竞赛预算：27 000 欧元（奖金 22 000 欧元，荣誉奖 5000 欧元）

评委
建 筑 评 委：Prof. Dr. Martina Abri, Architect, Potsdam/Berlin; Prof. Donatella Fioretti, Architect, Berlin; Hanna Gläsmer, Architect, Landkreis Oder-Spree; Sirko Hellwig, Architect, Guben; Prof. Manfred Ortner, Architect, Berlin/Vienna/Düsseldorf; Dr. Andrzej Poniewierka, Architect, Wroclaw; Prof. Henning Rambow, Architect, Leipzig
专 家 评 委：Fritz Taschenberger, Former Mayor of the City of Beeskow; Siegfried Busse, City Council Chairman, Bürgerverband Parliamentary Group; Eberhard Birnack, City Councillor, CDU Parliamentary Group Chairman; Dr. Karin Niederstraßer, City Councillor, Die LINKE Parliamentary Group Chairman; Sieghard Scholz, City Councillor, SPD Parliamentary Group Chairman; Hartmut Rudolph, City Councillor, FDP/Bauernverband Parliamentary Group Chairman

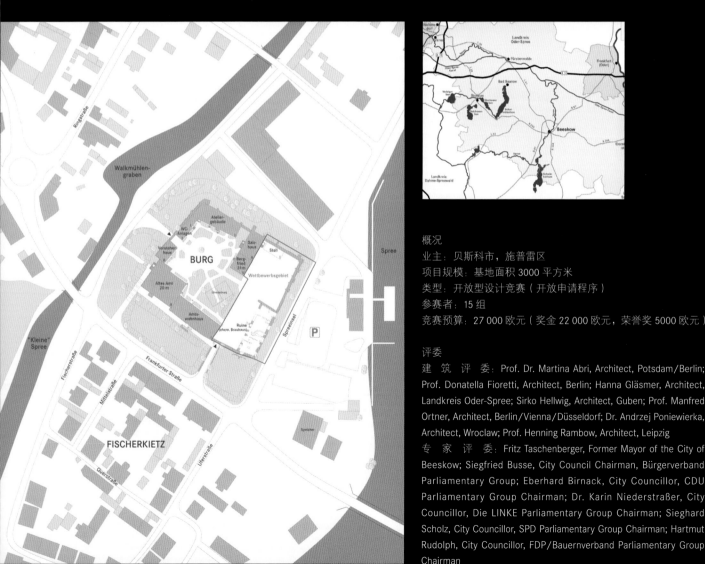

"项目目标一方面是为两个国家艺术收藏馆（原贝斯科艺术档案馆和柏林艺术家社会进步社）创建共同的档案保存处，另一方面是在贝斯科城堡东部的"开放带"上建造一个"闭合带"。档案馆的建筑设计应该尊重基地现有条件，与贝斯科城堡相融合，成为一个兼容的集合体，并履行相关建筑展览和艺术归档的职责。"

——引自《竞赛摘要》

新贝斯科艺术档案馆

项目进行之初，原贝斯科艺术档案馆和柏林艺术家社会进步社合并为新贝斯科艺术档案馆，作为艺术藏品的永久保地。原贝斯科艺术档案馆是德意志民主共和国的一个视觉艺术文献中心。在民主德国解体后，23 000 份原本归各个政治党派和社团组织所有的艺术藏品被移至公共领域。得益于1950—2003年间政府艺术基金的资助，柏林艺术家社会进步社持有约 14 000 份艺术藏品。这些艺术藏品见证了德国历史文化的进步与发展。新贝斯科艺术档案馆是德国历史及文化艺术的学习和研究中心，也是文化景观和施普雷区的旅游景点。

贝斯科城堡

竞赛场地位于德国勃兰登堡贝斯科东部历史城镇中心边缘的施普雷河分叉处。贝斯科城堡历史悠久，起源于 13 世纪，如今作为该地区的教育文化中心和音乐学校。场地中心是一座始建于中世纪的博物馆及地下室和其他附属展览空间。新建筑在一定程度上整合了前啤酒厂和东部城堡旧墙遗址。

竞赛任务

新贝斯科艺术档案馆的面积超过 3000 平方米，彼此相邻的小房间和小型公共展览区主要承接在贝斯科城堡内举办的一些小型活动，如艺术展览等。除了艺术藏品的保存，新贝斯科艺术档案馆还经常组织举办一些科学技术和文化艺术方面的国家级大型专题展览。此外，方案设计需就其本身如何尊重并呼应周边环境提出相应的建议，并以该背景中的物质空间和色彩关系作为参考，同时强调建筑的独立性和创新性。

新贝斯科艺术档案馆，贝斯科

参赛者

获奖者

1 一等奖 Max Dudler Architekten, Berlin (D)
2 二等奖 Marte.Marte Architekten, Weiler (D)
3 三等奖 CO A. Architektenkooperative, Berlin (D)
4 四等奖 Staab Architekten, Berlin (D)

荣誉奖

5 Peter Kulka Architektur, Dresden (D)
6 Stephan Braunfels Architekten, Berlin (D)
7 Nieto Sobejano Arquitectos, Madrid (E)
8 Kraaijvanger • Urbis, Rotterdam (NL)

第一轮

9 ARCHEA – MALE, Berlin (D)/Florence (I)
10 HG Merz, Berlin (D)
11 heneghan.peng architects, Dublin (IRL)
12 ahrens grabenhorst architekten, Hanover (D)
13 Architekt Johannes Walther, Hamburg (D)
14 lauth : van holst architekten, Wiesbaden (D)
15 Brunhart Brunner Kranz Architekten, Balzers (FL)

7

8

9

10

11

12

13

14

15

竞赛一等奖　获奖事务所

Max Dudler Architekten

Berlin (D)

方案设计者： Max Dudler
参与者： Sebastian Jonas Wolf, Maike Schrader
顾问： LAP, Berlin (D); Heimann Ingenieure, Berlin (D); Müller BBM, Berlin (D); HHP, Berlin (D); Silke Langenberg, Zurich (CH), Modellbau Milde, Berlin (D)

NEUES KUNSTARCHIV BEESKOW

Max Dudler Architekten

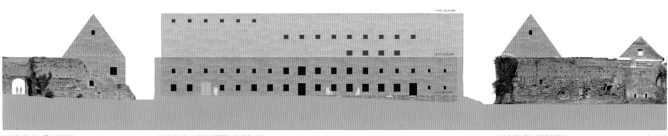

ANSICHT VON SÜDWESTEN ANSICHT VON NORDWESTEN (BURGHOF) ANSICHT VON NORDOSTEN M 1:200

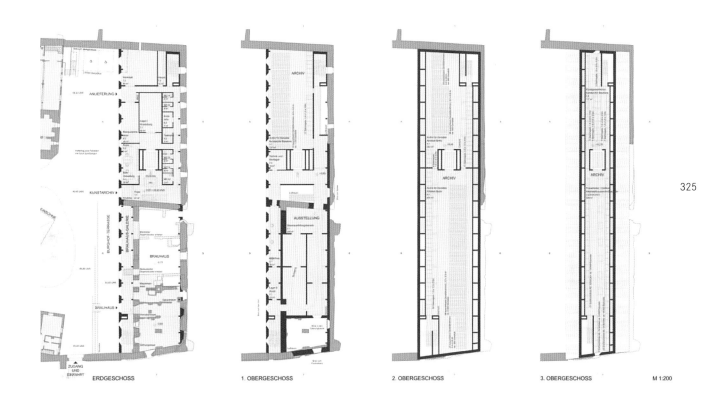

ERDGESCHOSS 1. OBERGESCHOSS 2. OBERGESCHOSS 3. OBERGESCHOSS M 1:200

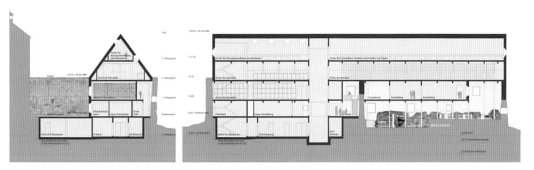

QUERSCHNITT A LÄNGSSCHNITT 1

UNTERGESCHOSS

新贝斯科艺术档案馆，贝斯科

竞赛二等奖　获奖事务所

Marte.Marte Architekten

Weiler (A)

方案设计者：Bernhard Marte, Stefan Marte
参与者：Bettina Tóth, Timo Bereiter, Robert Zimmermann, Marcus Jung

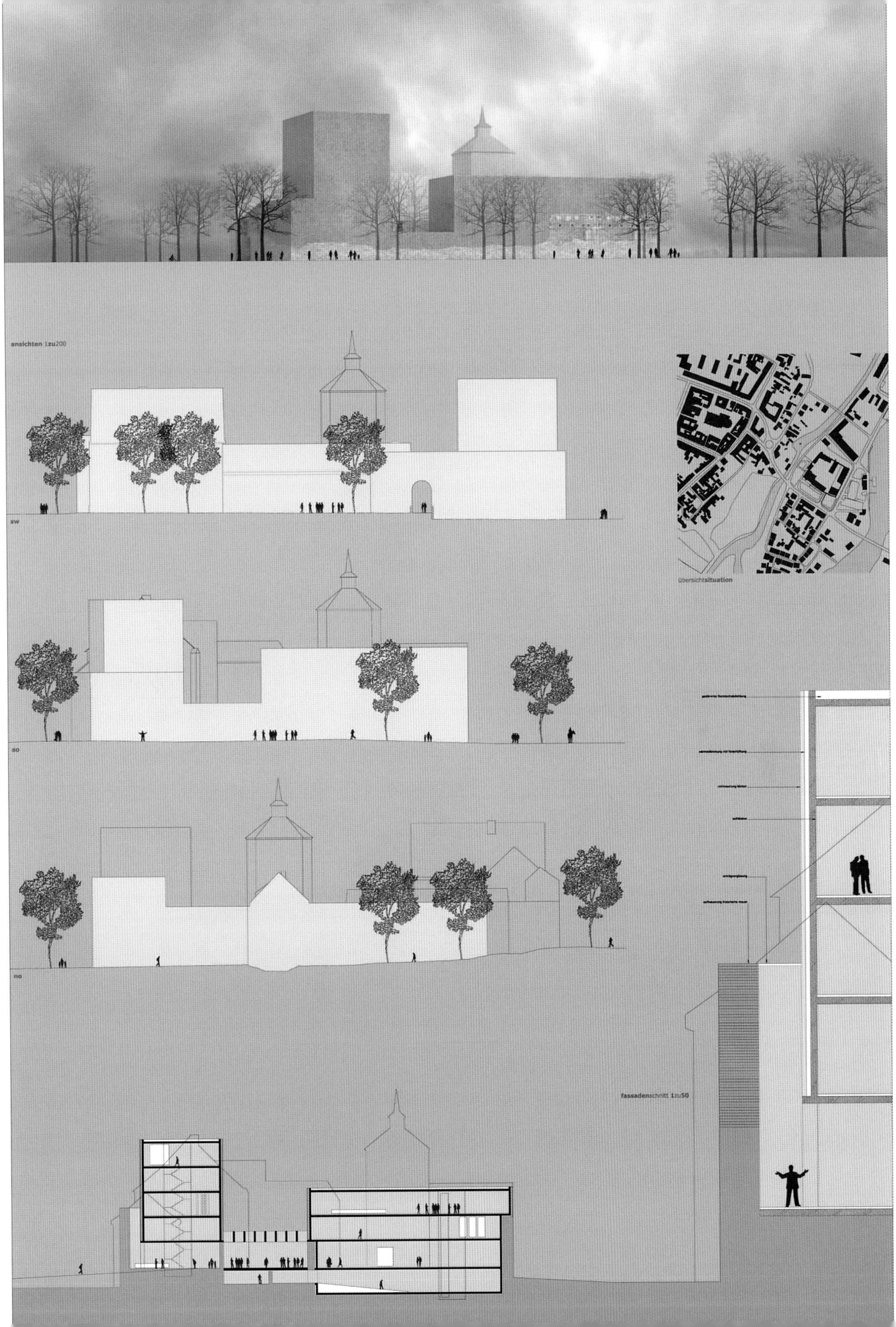

ensichten 1zu200

sw

so

no

übersichtsituation

fassadenschnitt 1zu50

竞赛三等奖　获奖事务所

CO A. Architektenkooperative

Berlin (D)

方案设计者： Jakob Koenig
参与者： Mathias Kl.pfel, Christian Kahl
顾问： DBV - Dierks, Babilon und Voigt, Berlin (D); BLS Energieplan, Berlin (D), Mathias Ludwig

Erdgeschoss M 1:200   1. Obergeschoss M 1:200

Querschnitt M 1:200   Längsschnitt M 1:200

Ansicht Hofseite M 1:200   Ansicht Nordseite M 1:200

Ansicht Südseite M 1:200   Ansicht Ostseite M 1:200

竞赛四等奖　获奖事务所

Staab Architekten

Berlin (D)

方案设计者：Volker Staab
参与者：Petra W.Idle, Bettina Schriewer, Max Illing, Ivan Kaleov
顾问：iKM - Ingenieurbüro Kless Müller, Dresden (D), Rainer Kless

Neues Kunstarchiv Beeskow

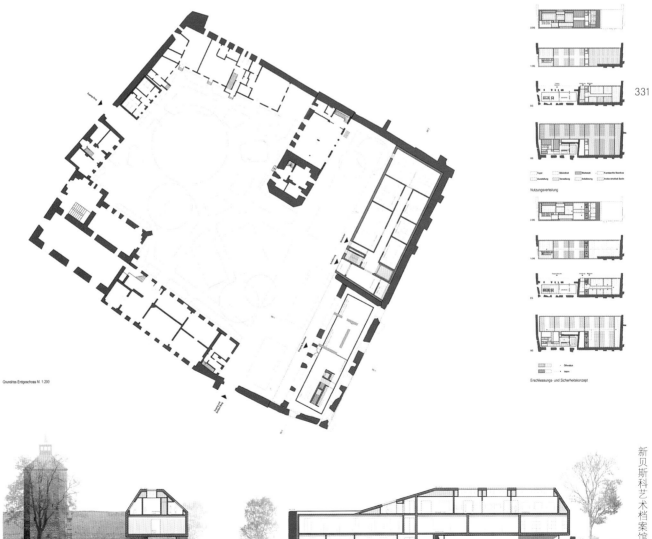

莫斯科理工博物馆

俄罗斯 莫斯科

2012—2013

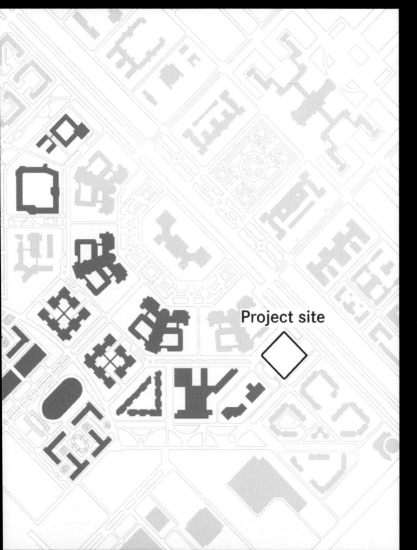

概况
业主：莫斯科理工博物馆基金会
竞赛管理：斯特列尔卡学院与 [phase eins]. 工作室联合设计，斯特列尔卡学院提供多媒体支持
项目规模：基地面积 35 000 平方米
类型：双阶段设计竞赛
参赛者：6 组
竞赛预算：180 000 欧元

评委
James Alexander, CEO at Event Communications Ltd., London; Andrey Vladimirovich Bokov, General Director of the Moscow State Unitary Enterprise "Moscow Research and Development Institute for the Culture, Recreation, Health and Sport Facilities 'Mosprojekt-4'", President of Russia's Union of Architects; Caroline Bos, Urban Planner, UNStudio, Amsterdam; Andrey Evgenievich Busygin, Deputy Minister of Culture of the Russian Federation; Robert Firmhofer, Director of the Copernicus Science Centre, Warsaw and President of the Ecsite-European Network of Science Centres and Museums; Jan Kleihues, Architect; Sergey Olegovich Kuznetsov, Chief Architect of Moscow and First Deputy Chair of the Moscow Architecture and Urban Development Committee; Nikolay Alexandrovich Novikov, Head of the Capital Construction Division at Lomonosov Moscow State University; Grigory Isaakavoch Revzin, Architecture Critic and Publicist, Moscow; Julia Vasilyevna Shakhnovskaya, Deputy General Director of the Polytechnic Museum, Moscow

"莫斯科理工博物馆及其附属教育中心的建造将有助于莫斯科国立大学校园的景观营造,它有望成为极具吸引力且令人流连忘返的地方。"

——引自《竞赛摘要》

莫斯科理工博物馆
莫斯科理工博物馆是世界上历史最悠久、规模最大的科技博物馆之一。该项目最重要的部分(重新组织并更新市中心的历史建筑)于2010年启动。博物馆位于莫斯科国立大学校园(俄罗斯最大的公立大学)内的新增建设地点。项目目标是建造一座当代博物馆,用于展示先进的科学技术和多媒体技术以及提供公共艺术教育。除了供莫斯科大学的学生与教职员工使用,博物馆平时对公众开放,旨在丰富市民的文化生活,并为国内外专家学者的技术交流提供平台。每到周末,很多家长带着孩子来到博物馆参观、学习,孩子们被带入科学展览、多媒体技术、电影制作等高科技和艺术世界中,这对丰富知识体系和拓宽视野具有极大的帮助。

莫斯科国立大学
莫斯科国立大学位于市中心的西部边缘,周围是一个公园。宽阔的街道和纪念性建筑穿过校园。校园里最著名的建筑是于1953年设计的古典主义风格 Lomonosovsky 大街主建筑。253 米高的塔楼坐落于莫斯科市中心,是"七姐妹"环形高楼的其中之一。市民在此可观看比赛。

竞赛任务
规划区域约 32 000 平方米,包括实验室、临时展览区、永久展览区、演讲厅、会议室、图书馆、科学/艺术馆、电影放映室、多媒体中心以及一个大型中央大厅。科学/艺术馆经常举办一些高新技术展览,电影放映室经常介绍有关电影制作的知识。从城市规划的角度分析,博物馆应该融入校园的整体结构,并成为一座里程碑式建筑。此外,该项目的附属工程是面朝 Lomonosovsky 大街的广场设计和面向博物馆主建筑的视觉轴线设计。

参赛者

1 获奖者 MASSIMILIANO FUKSAS Architetto, Rome (I) with SPEECH, Moscow (RUS)

决赛者

2 Project Meganom, Moscow (RUS), with John McAslan + Partners, London (GB)
3 3XN A/S, Copenhagen (DK), with ASADOV Architectural Studio, Moscow (RUS)
4 MECANOO International, Delft (NL), with REZERVE, Moscow (RUS)
5 FARSHID MOUSSAVI ARCHITECTURE, London (GB), with AB Rozhdestvenka, Moscow (RUS)
6 LEESER Architecture, Brooklyn (USA) with ABD Architects, Moscow (RUS)

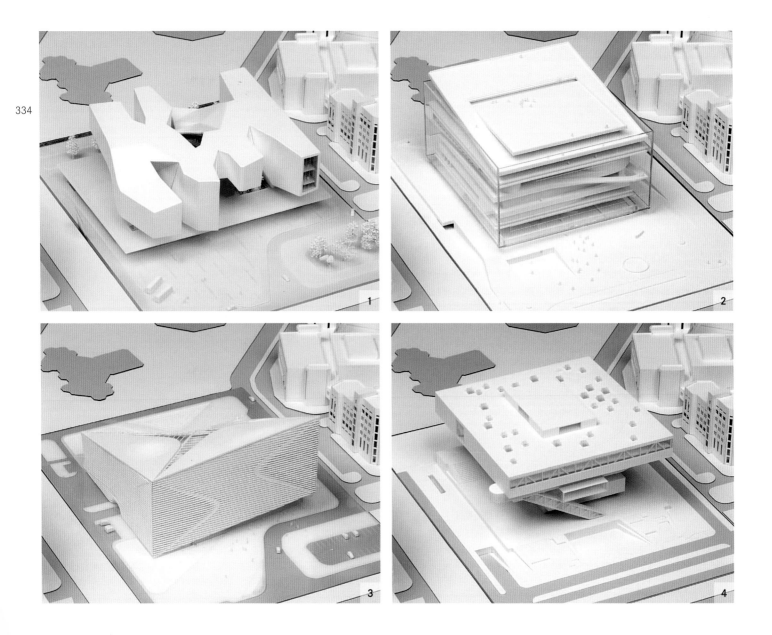

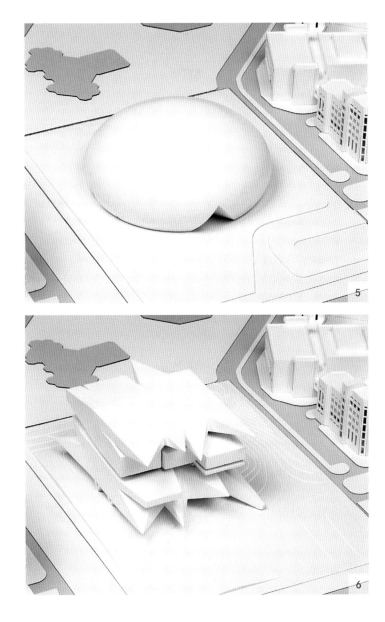

参赛者

莫斯科理工博物馆，莫斯科

获奖者 获奖事务所

MASSIMILIANO FUKSAS Architetto

Rome (I); with SPEECH, Moscow (RUS)

方案设计者：Massimiliano Fuksas, M.H. Desyatnikov

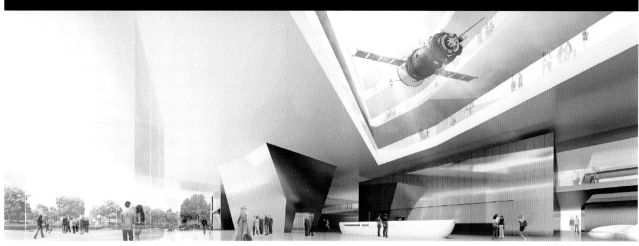
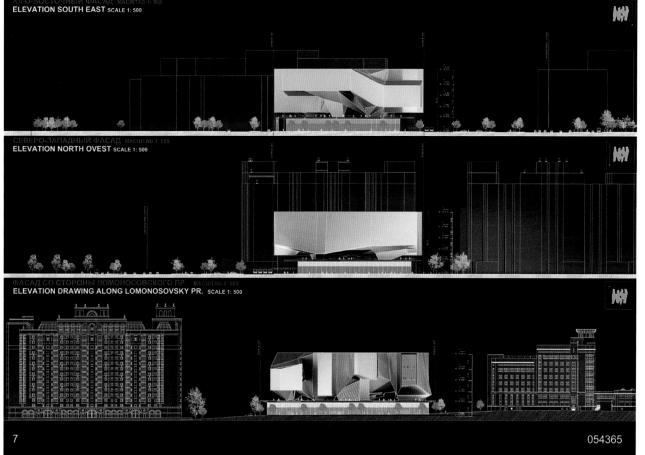

决赛者 获奖事务所

Project Meganom

Moscow (RUS); with John McAslan + Partners, London (GB)

方案设计者：Grigoryan Yury, Aidan Pottee

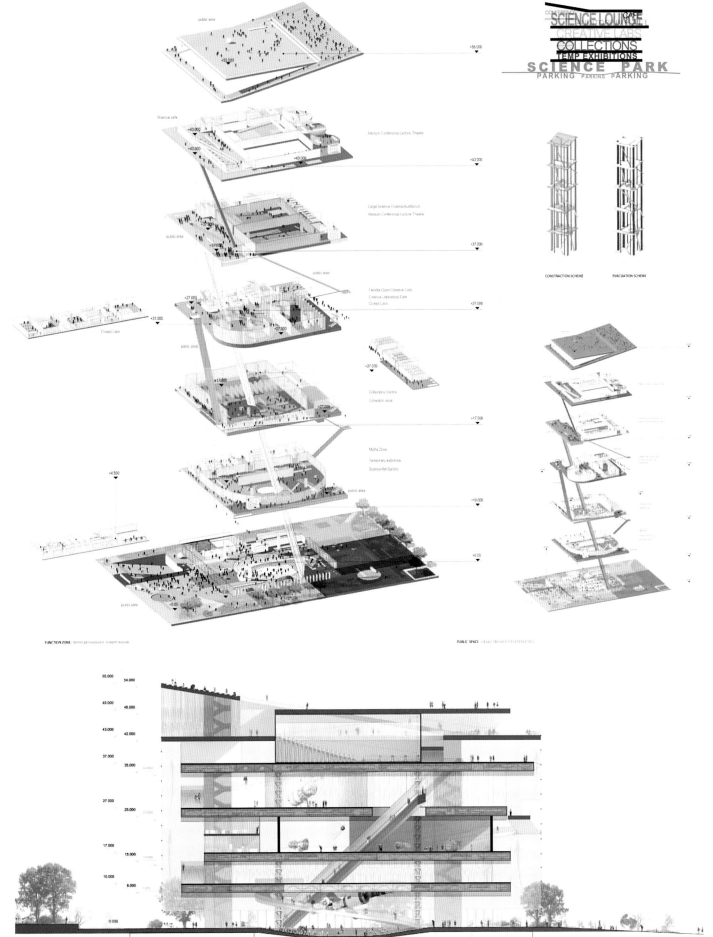

决赛者 获奖事务所

3XN A/S

Copenhagen (DK); with ASADOV Architectural Studio, Moscow (RUS)

方案设计者：Kim Herforth Nielsen, Andrey Alexandrovich Asadov

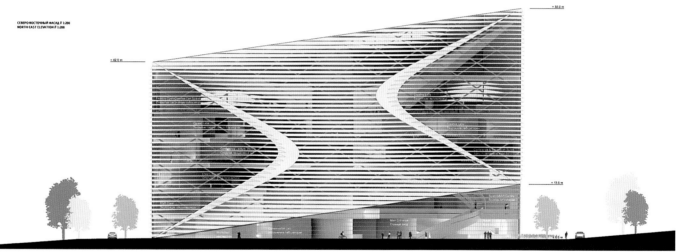

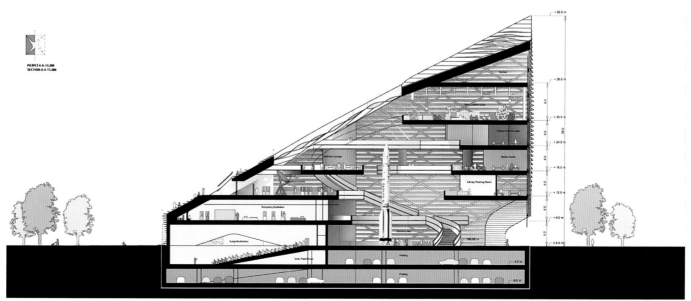

决赛者　获奖事务所

MECANOO International

Delft (NL); with REZERVE, Moscow (RUS)

方案设计者：Aart Fransen, Semen Elimovich Lamdon

Перспективные изображения, общие виды здания
External and internal visualisations

Развертка по Ломоносовскому проспекту   масштаб 1:500
Elevation drawing along Lomonosovsky Avenue   scale 1:500

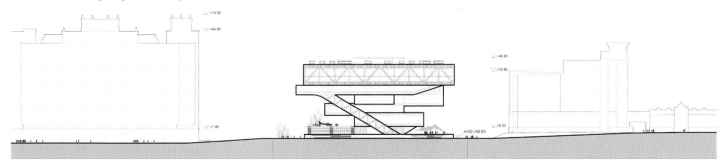

Фасады 2, 3 и 4   масштаб 1:500
Elevations 2, 3 & 4   scale 1:500

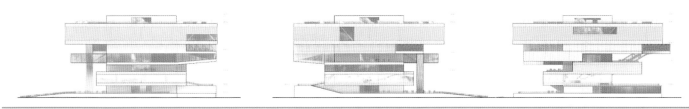

决赛者 获奖事务所

FARSHID MOUSSAVI ARCHITECTURE

London (GB); with AB Rozhdestvenka, Moscow (RUS)

方案设计者：Farshid Moussavi, Narine Tutcheva

Западный фасад / West elevation 1:200

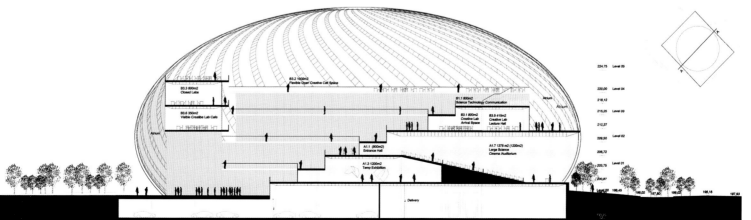

Разрез / Section A-A 1:200

决赛者 获奖事务所

LEESER Architecture

Brooklyn (USA); with ABD Architects, Moscow (RUS)

方案设计者: Thomas Leeser, Boris Levyant

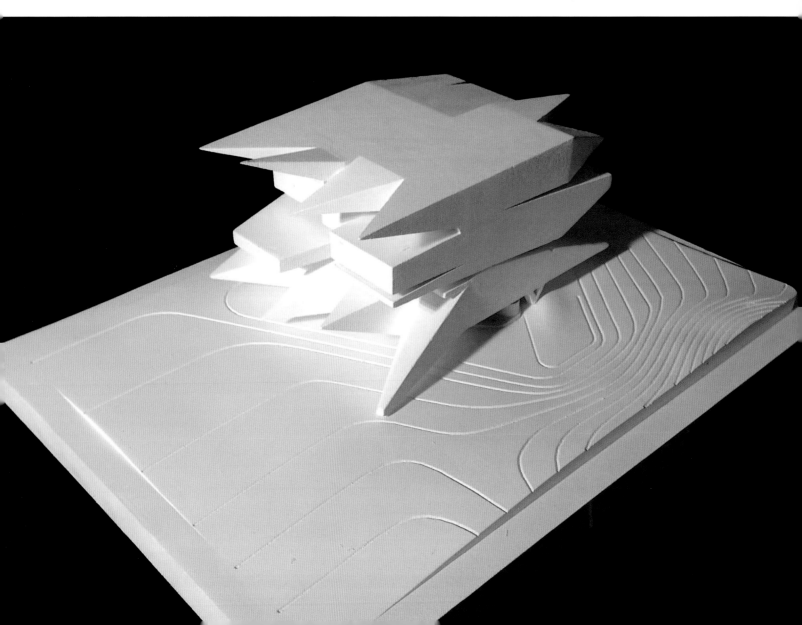

PROGRAM CONCEPT

Концепция здания Музейно-просветительского центра построена на программной двойственности в отношениях между Посетителями и Исследователями. Переплетение двух программных рядов не только создает заданный алгоритм в структуре здания, но и буквально организовано как двойная спираль биологической цепи ДНК. Такая организация создает максимально интенсивный, наиболее функциональный и богатый пространственный опыт для Исследователей и Посетителей, предполагая практически неограниченное взаимодействие между ними: в любой момент существует возможность не только увидеть пространства противоположной программы (например, из пространства посетителей в пространство исследователей, и наоборот), но и создать контролируемые переходы и доступ между ними.

The Polytechnic Education Center is build around the programmatic duality and relationship of User and Visitor. A helical relationship of the two program sequences not only create the fundamental DNA of the building, but is literally organized like the double helix of a biological DNA strand. This organization creates the most intense, most functional and most enriching experiences for Users and Visitors, while allowing an almost infinite flexible relationship between the two.

At any given point the possibility exists to not only see into the opposing program spaces (eg. from Visitor space to User space and vice versa), but to allow controlled shortcuts and access between them.

Mystetskyj 兵工厂文化区

乌克兰 基辅

2007—2008

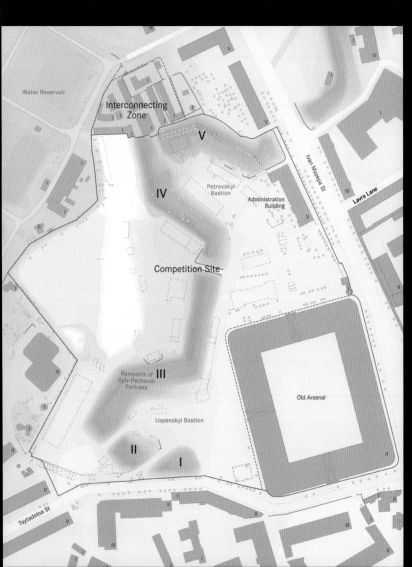

概况

业主：乌克兰国家事务部

项目规模：基地面积约 5000 平方米

类型：双阶段开放型设计竞赛（开放申请程序）

参赛者：第一阶段 21 组，第二阶段 7 组

竞赛预算：450 000 欧元（第二阶段奖金 170 000 欧元；第二阶段费用 40 000 欧元）

评委

Igor Tarasyuk, Head of State Department of Affairs; Vasyl Vovkun, Minister of Culture and Tourism; Ivan Vasyunyk, Vice Prime Minister of Ukraine; Yuriy Bogutsky, Deputy Head of the Secretariat of the President of Ukraine; Igor Didkovsky, Director General of State Enterprise Cultural-Art and Museum Complex "Mystetskyj Arsenal"; Igor Shpara, President of the Ukrainian Union of Architects; Vasyl Prysyazhnyuk, Main Architect of Kyiv; Louis Becker, Architect, Copenhagen; Prof. Rebecca Chestnutt, Architect, Berlin; Philippe Chaix, Architect, Paris; Prof. Marcel Meili, Architect, Zurich; Prof. Dr. Vladimir ˇSlapeta, Architect, Brno/Prague; Prof. Guenther Vogt, Landscape Architect, Zurich

"参赛者面临的挑战是打造一片新颖且独具特色的城市区域，将多种功能嵌入其中，并使其和谐地融入周边环境。该区域以基辅佩彻斯克'洞穴'修道院、基辅佩彻斯克要塞和Mystetskyi兵工厂老建筑作为经典建筑地标。"

——引自《竞赛摘要》

基辅佩彻斯克

10公顷的竞赛场地位于基辅佩彻斯克区域，毗邻历史悠久的第聂伯河西部河岸。该区域中最古老的建筑是名为"洞穴"的修道院（Pecherska lavra，乌克兰语：洞穴），直至今日，它一直是东正教的精神中心。"洞穴"修道院与圣索菲亚大教堂一同被联合国教科文组织列为"世界文化遗产地"。Mystetskyi兵工厂始建于19世纪，位于基辅佩彻斯克要塞区域内，也是Mystetskyj兵工厂文化区的中心所在。

Mystetskyi兵工厂

今天的Mystetskyi兵工厂，其建筑的文化意义远远胜过军事意义。兵工厂建筑致力于增强乌克兰人民的文化认同感，促进跨国界的建筑文化影响及艺术交流，成为欧洲顶级建筑艺术的目的地，在设计理念及功能结构方面媲美国际同类建筑，巩固乌克兰建筑在欧洲建筑领域中的地位。

竞赛任务

竞赛任务包括平衡城市框架和基地的复杂关系，建造几座国家级的文化建筑，兵工厂和城墙被整合到展览空间中。60 000平方米的规划区域用于建造乌克兰音乐中心、现代美术馆、国际电影中心、康复中心、酒店以及行政办公楼。总的来说，Mystetskyi兵工厂作为该区域中最突出的建筑地标。建筑设计与景观营造既承接历史，又反映现代。然而，基于2008年后出现的经济危机及随之而来的一系列政治变革，该项目在具体实施过程中相比较竞赛中确定的方案设计做了一些显著的更改与变动。

第一阶段参赛者

第二阶段入围者

1　David Chipperfield Architects, Berlin (D)
2　Arata Isozaki & Associates, Tokyo (J)
3　AIX Arkitekter, Stockholm (S)
4　Schmidt Hammer Lassen Architects, Copenhagen (DK)
5　Kleihues + Kleihues Architekten, Berlin (D)
6　Schweger Assoziierte Gesamtplanung, Hamburg (D)
7　Dietrich | Untertrifaller Architekten, Bregenz (A)

其他参赛者

8 Arkkitehtitoimisto Lahdelma & Mahlamäki, Helsinki (FIN) 9 Dominique Perrault Architecture, Paris (F) 10 Architecture Studio V. Schevchenko, Kiev (UA) 11 agps architecture, Los Angeles (USA) 12 Architectural Bureau Zotov & Co., Kiev (UA) 13 Paulo David Arquitecto, Madeira (P) 14 Willmotte & Associés, Paris (F) 15 Claus en Kaan Architecten, Rotterdam (NL) 16 Sadar Vuga Arhitekti, Ljubljana (SLO) 17 Asymptote Architecture, New York (USA) 18 Klaus Kada, Graz (A) 19 Sauerbruch Hutton Architekten, Berlin (D) 20 Ivan Kroupa Architekti, Prague (CZ) 21 Adjaye Associates, London (GB)

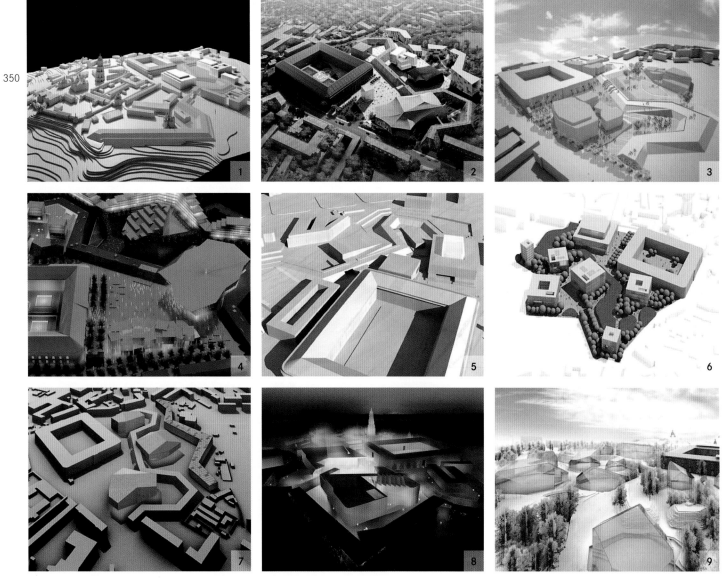

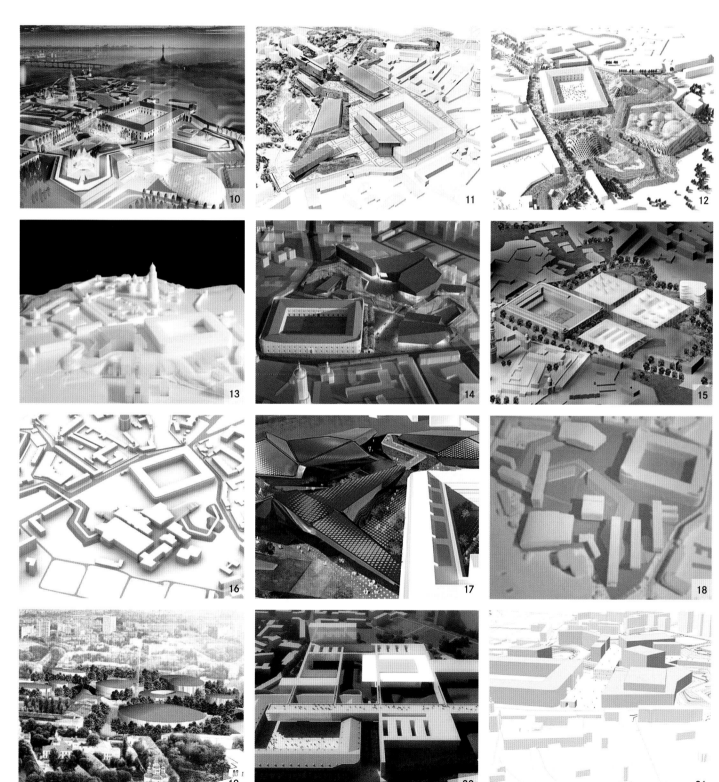

第二阶段参赛者

获奖者

1 一等奖 Arata Isozaki & Associates, Tokyo (J)
2 二等奖 AIX Arkitekter, Stockholm (S)
3 三等奖 David Chipperfield Architects, Berlin (D)
4 四等奖 Kleihues + Kleihues Architekten, Berlin (D)

第二轮

5 Schmidt Hammer Lassen Architects, Copenhagen (DK)

第一轮

6 Dietrich | Untertrifaller Architekten, Bregenz (A)
7 Schweger Assoziierte Gesamtplanung, Hamburg (D)

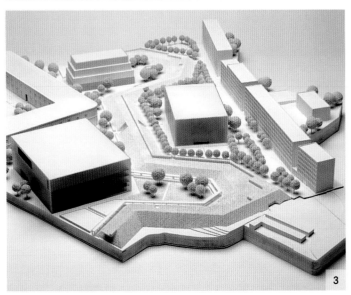

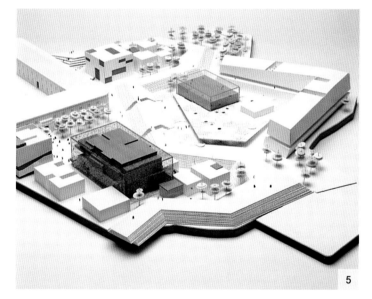

竞赛一等奖　获奖事务所

Arata Isozaki & Associates

Tokyo (J)

方案设计者： Arata Isozaki
参与者： Hiroshi Aoki, Ryosuke Kamano, Hiroshi Yoshino, Takayuki Uchida, Yoko Sano, Kaori Moriyama, Mitsuyoshi Miyazaki, Umeoka Koji, Ana Martins, Kai Kasugai
自由职业者： Ingarden & Ewy - Architekcki, Krakow (PL), Krysztof Ingarden, Piotr Urbanowicz
顾问： Ove Arup, London (GB)

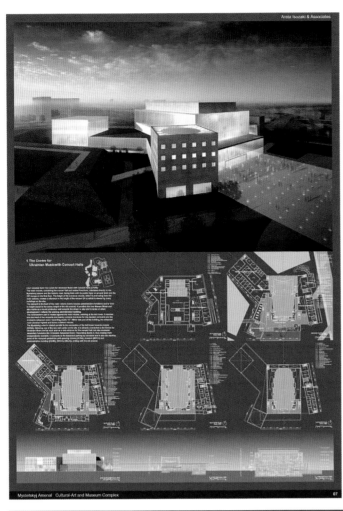

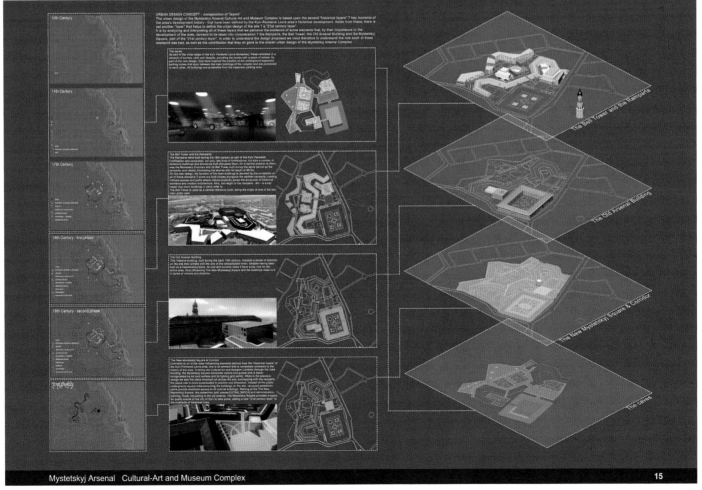

Mystetskyj Arsenal Cultural-Art and Museum Complex

竞赛二等奖　获奖事务所

AIX Arkitekter

Stockholm (S)

方案设计者：Lars Johansson, Jesper Engstr.m
参与者：Yvan Ikhlef, Lisa Wikstr.m, Anders Scherman, Eva G.ransson, Torsten Nobling, Harald Keijer, Peder Lindbom, Lotta Lindgren, Marcus Hahn, Fredrik Ekerhult, Frida Rosenberg, Erik K.llstr.m
自由职业者：Nyrens Arkitektkontor, Stockholm (S), Pia Englund; Sweco Architects, Stockholm (S), Jimmy Norrman; Sweco Infrastructure, Stockholm (S), Sverker Hanson
顾问：Jan Cassel, Stockholm (S)

MYSTETSKYJ ARSENAL
CULTURAL-, ART- AND MUSEUM COMPLEX

Center for Ukrainian Music with Concert Halls

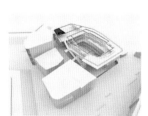

Distribution of uses as per task, colour as in spatial program
september 12 2008

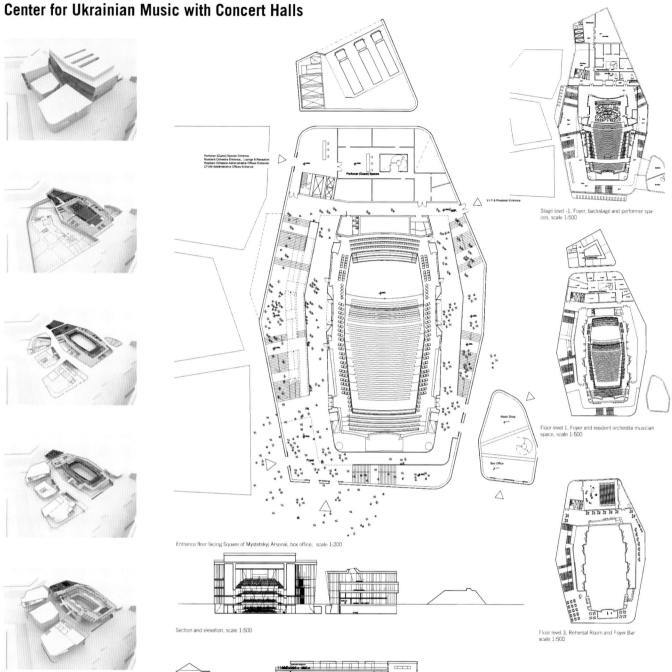

Stage level -1, Foyer, backstage and performer spaces, scale 1:500

Floor level 1, Foyer and resident orchestra musician space, scale 1:500

Floor level 3, Rehersal Room and Foyer Bar scale 1:500

Entrance floor facing Square of Mystetskyj Arsenal, box office, scale 1:200

Section and elevation, scale 1:500

Longitudinal section through Concert Hall and Rehersal Room, possible to use as secondary concert hall for chamber music, scale 1:500

View over Lavra Cloister Area from CFUM top floor restaurant and bar

竞赛三等奖　获奖事务所

David Chipperfield Architects

Berlin (D)

方案设计者： Mark Randel (Design Director), Harald Müller (Managing Director)
参与者： Cyril Kriwan (Project Architect), Andrea Garcia Crespo, Ivan Dimitrov, Ulrike Eberhardt, Sebastian Epkes, Annette Flohrschütz, Pavel Frank, Gesche Gerber, Fernando Gomez Martinez, Dirk Gschwind, Michael Haverland, Soeren Johansen, Mikhail Kornev, Thomas Kupke, Dalia Liksaite, Marcus Mathias, Lijun Shen, Ute Zscharnt
自由职业者： Fernando González; Imaging Atelier, Helsinki (FIN), Jens Gehrken
顾问： Wirtz International, Schoten (B); Ingenieurgruppe Bauen, Berlin (D)

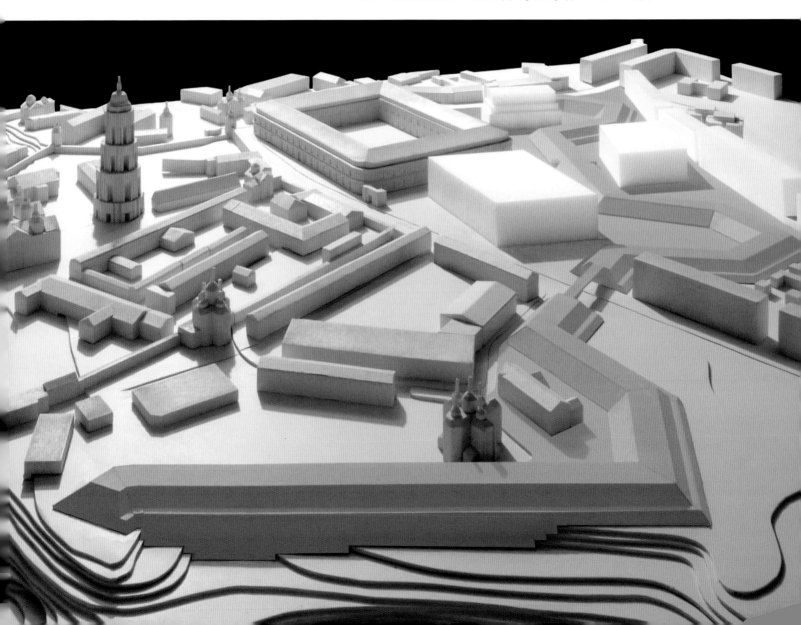

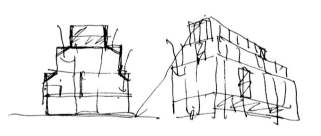
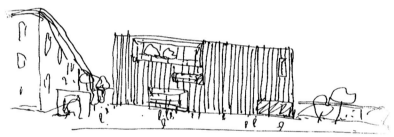

竞赛四等奖 获奖事务所

Kleihues + Kleihues Architekten

Berlin (D)

方案设计者： Jan Kleihues, Klaus Schuwerk
自由职业者： Paolo Casaburi, Claudia Sasso, Antonella Capozzi, Rosella Cincotti, Johannes Kressner, Mariachiara Baldassare
顾问： Lützow 7: Cornelia Müller Jan Wehberg, Garten- und Landschaftsarchitekten, Berlin (D); Ove Arup & Partners Berlin (D); GRI Gesellschaft für Gesamtverkehrsplanung, Regionalisierung und Infrastrukturplanung, Berlin (D), Bodo Fuhrmann

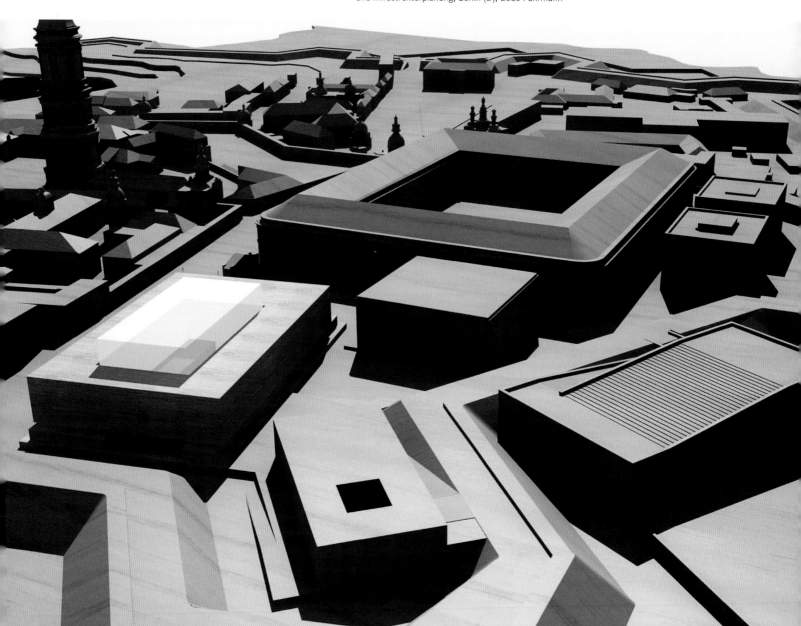

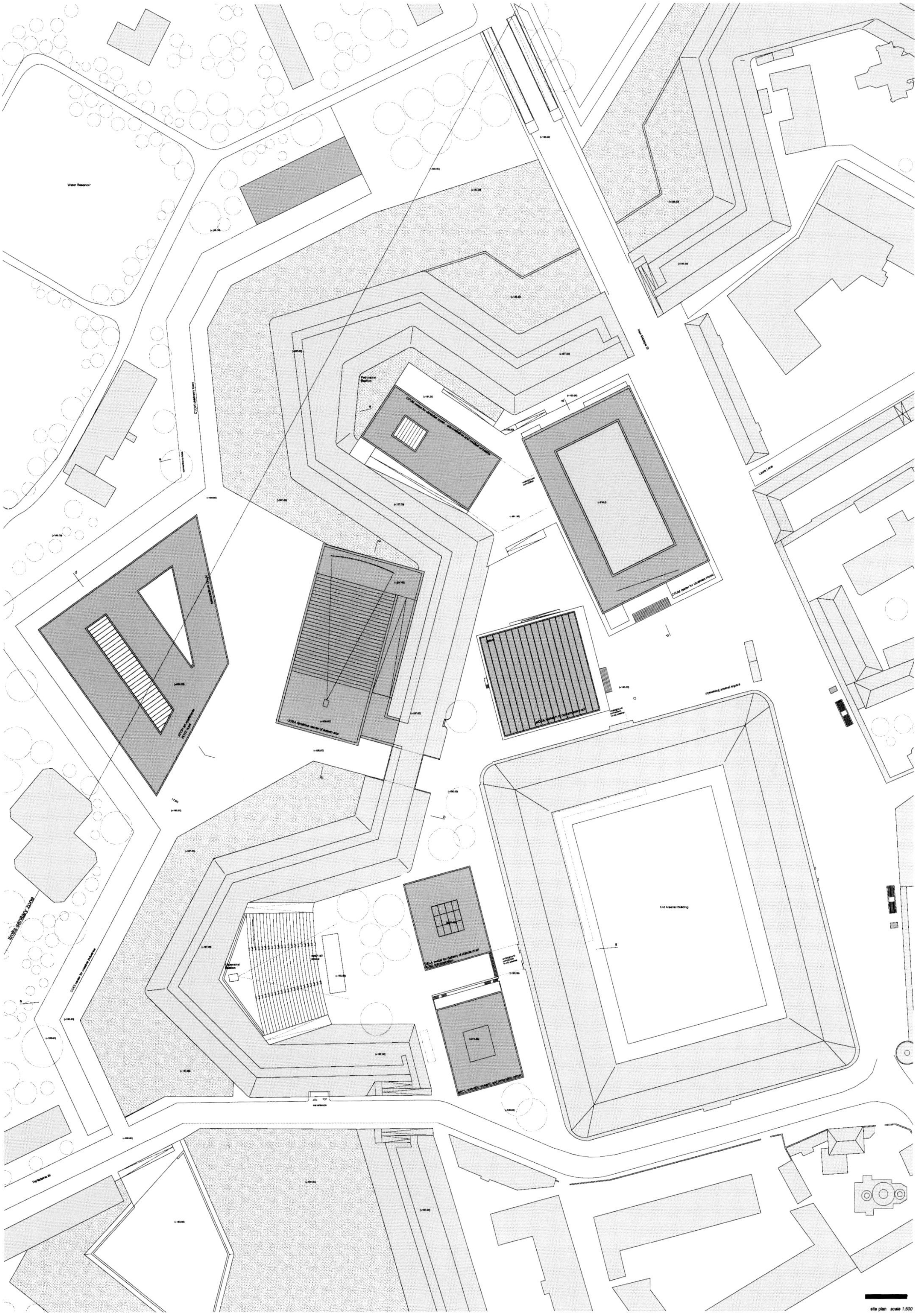

Mystetskyj 兵工厂文化区，基辅

Kleihues + Kleihues Architekten

kleihues + schuwerk

site plan scale 1:500

洲际酒店、滑冰俱乐部、音乐厅

奥地利 维也纳

2013—2014

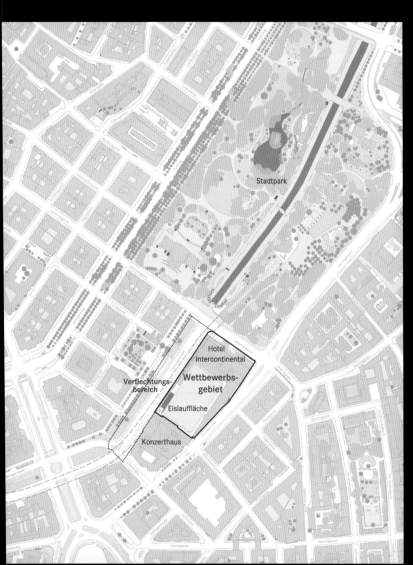

概况

业主：WertInvest Hotelbeteiligungs 电子商务有限公司
项目规模：基地面积约 60 000 平方米
类型：双阶段开放型设计竞赛（开放申请程序）
参赛者：第一阶段 24 组，第二阶段 6 组
竞赛预算：324 000 欧元（第一阶段费用 6000 欧元，第二阶段费用 15 000 欧元，第二阶段奖金 90 000 欧元）

评委

建 筑 评 委：Thomas Madreiter, Planning Director of Vienna; Franz Kobermaier, Head of Municipal Department 19 of Vienna; Erich Steinmayr, Architect, Feldkirch; Rainer Köberl, Architect, Innsbruck; Prof. Markus Allmann, Architect, Munich; Kai-Uwe Bergmann, Architect, Copenhagen; Guido Hager, Landscape Architect, Zurich; Prof. Wilfried Kühn, Architect, Berlin; Prof. Regine Leibinger, Architect, Berlin; Prof. Rudolf Scheuvens, City Planner Dortmund/Vienna

专 家 评 委：Rudolf Zabrana, Deputy Director, Municipal Office for the 3rd District, Vienna; Dr. Michael Tojner, Vorstand, WertInvest; Christoph Chorherr, Member of the Vienna City Council

"维也纳城市中心的都市与建筑品质是价值不断变化的突出例证。"
——引自《竞赛摘要》

维也纳环形路
三座建筑定义了维也纳环形路,并使其在城市布局中脱颖而出:洲际酒店、滑冰俱乐部室内外溜冰场、音乐厅。项目目标是打造一个城市综合体,整合商务会议、运动休闲、音乐欣赏等多种功能,营造健康、舒适、愉悦的环境氛围,满足市民和游客的使用需求。

场地
竞赛场地约 15 400 平方米,位于维也纳市中心的城市公园和施瓦岑贝格广场之间。该区域位于历史缓冲区(16 世纪的防御工事)和被联合国教科文组织列为"世界文化遗产地"的历史核心区的外围。除了竞赛区域,城市规划建议在 Lothringer 大街 10 000 平方米的范围内建造一系列建筑。现有建筑布局对该区域与城市整体结构之间的有机关联造成一定的阻碍,且前往邻近区域的可达性较差。

竞赛任务
规划区域包括 450~500 间酒店客房、会议设施、餐厅以及 50 套公寓。滑冰俱乐部包括 6000 平方米的室外溜冰场、1800 平方米的室内溜冰场以及其他附属空间。其中一项竞赛任务是参赛者讨论在遵循"城市适应性"原则的前提下如何处理现有 44.5 米高的酒店,或保存,或改造,或拆除,并得出结论。如果将其保存或改造,那么参赛者面临的一个特殊的挑战是如何在"世界文化遗产地"的背景下对酒店高度进行合理化控制。

洲际酒店、滑冰俱乐部、音乐厅,维也纳

第一阶段参赛者

第二阶段入围者

1. Isay Weinfeld Arquitectos e Urbanismo, São Paulo (BR)
2. Zeytinoglu Architects with LAND IN SICHT – Büro für Landschaftsplanung, Vienna (A)
3. Ortner & Ortner Baukunst, Vienna (A)
4. Atelier d'architecture Chaix & Morel et Associés, Paris (F), with DnD Landschaftsplanung, Vienna (A)
5. Max Dudler Architekten with Atelier Loidl Landschaftsarchitekten, Berlin (D)
6. querkraft architekten, Vienna (A)

其他参赛者

7 Machado and Silvetti Associates, Boston (USA), with Richard Burck Associates, Somerville (USA) **8** Snøhetta, Oslo (N) **9** Atelier Paolo Piva with Marid Terzic, Vienna (A) **10** ARSP, Dornbirn, with koeber Landschaftsarchitektur, Stuttgart (D) **11** Coop Himmelb(l)au, Vienna (A), with ST raum a. Landschaftsarchitekten, Berlin (D) **12** BWM Architekten und Partner, Vienna (A) with Wiel Arets Architects, Amsterdam (NL) and ARGE Standler/Zimmermann, Vienna (A) **13** 3XN A/S, Copenhagen (D), with WES LandschaftsArchitektur and Hans-Hermann Krafft, Berlin (D) **14** Berger + Parkkinen Architekten, Vienna (A), with Agence Ter.de Landschaftsarchitekten, Karlsruhe (D) **15** Sauerbruch Hutton Architekten with Sinai Landschaftsarchitekten, Berlin (D) **16** Freimüller Söllinger Architektur, Vienna (A), with Latz + Partner, Kranzberg (A) **17** Nieto Sobejano Arquitectos with Frank Kiessling Landschaftsarchitekten, Berlin (D) **18** C+S Architects, Treviso (I), with bauchplan).(, Munich (D) **19** Delugan Meissl Associated Architects, Vienna (A), with Burger Landschaftsarchitekten Partnerschaft, Munich (D) **20** Neutelings Riedijk Architecten with West 8 Urban Design & Landscape Architecture, Rotterdam (NL) **21** Arkkitehtitoimisto Lahdelma & Mahlamäki, Helsinki (FIN), with Masu Planning, Copenhagen (D) **22** all Design with LDA Design, London (GB) **23** Henke Schreieck Architekten with Auböck + Kárász Landschaftsarchitekten, Vienna (A) **24** Guillermo Vázquez Consuegra Arquitecto, Seville (E), with Arquitectura y Agronomia, Barcelona (E)

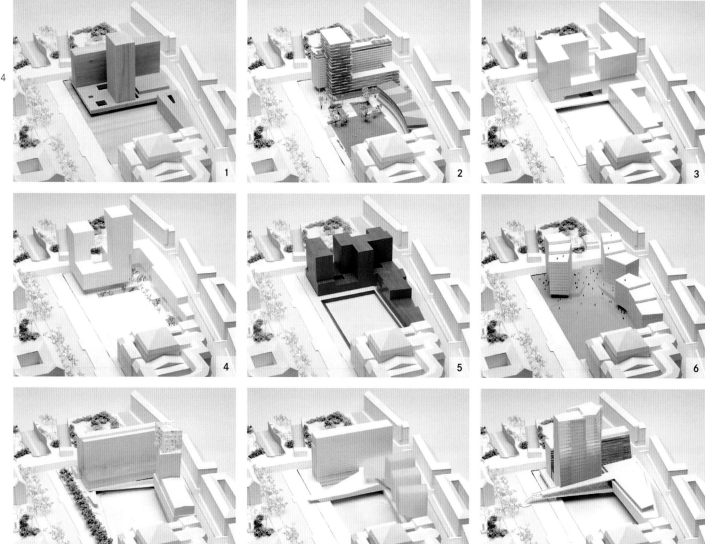

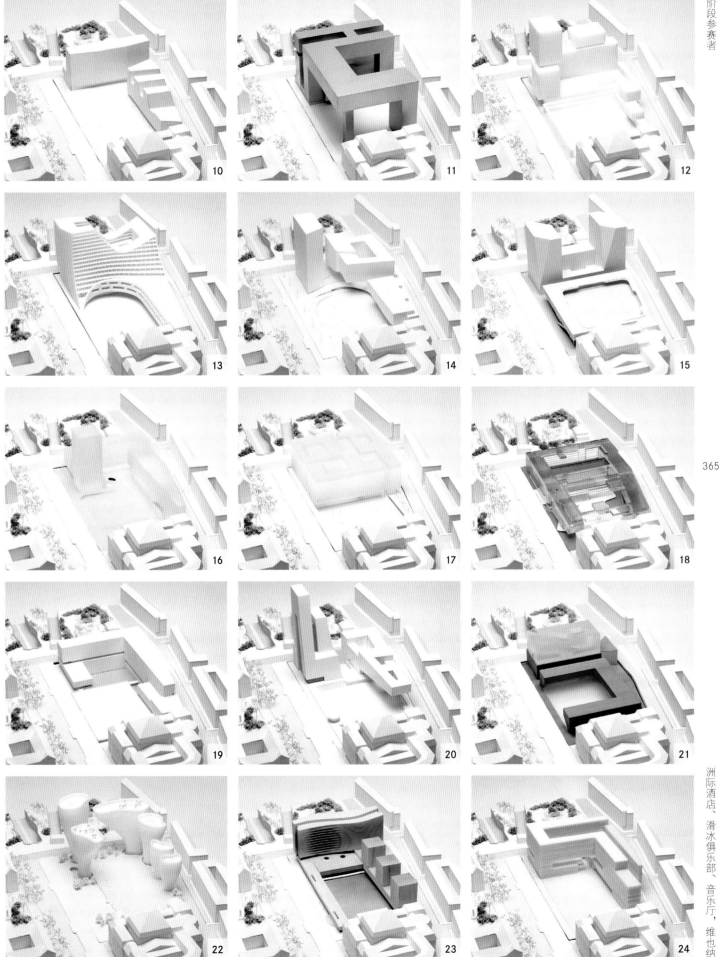

第二阶段参赛者

1 一等奖 Isay Weinfeld Arquitectos e Urbanismo, S.o Paulo (BR)
2 二等奖 Atelier d'architecture Chaix & Morel et Associés, Paris (F), with DnD Landschaftsplanung, Vienna (A)
3 荣誉奖 querkraft architekten, Vienna (A)

决赛者

4 Zeytinoglu Architects with LAND IN SICHT - Büro für Landschaftsplanung, Vienna (A)
5 Ortner & Ortner Baukunst, Vienna (A)
6 Max Dudler Architekten with Atelier Loidl Landschaftsarchitekten, Berlin (D)

竞赛一等奖　获奖事务所

Isay Weinfeld
Arquitectos e Urbanismo

São Paulo (BR)

方案设计者： Isay Weinfeld
参与者： Lucas Jimeno Dualde
自由职业者： Sebastian Murr, Katherina Deborah Ortner, Wolfram Winter
顾问： Bollinger, Grohmann und Schneider, Vienna (A); ZFG-Projekt, Baden (A); Bollinger, Grohmann Consulting, Frankfurt am Main (D); DBI - Düh Beratende Ingenieure, Vienna (A); Buttler Harrer Büro für Architektur, Vienna (A); Franziska Mayer-Fey, Herrsching (D)

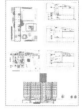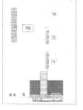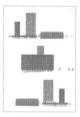

竞赛二等奖　获奖事务所

Atelier d'architecture Chaix & Morel et Associés

Paris (F); with DnD Landschaftsplanung, Vienna (A)

方案设计者：Jean-Paul Morel, Philippe Chaix, Walter Grasmug, Anna Detzlhofer, Sabine Dessovic
参与者：Remi Brabis, Benoit Chantelou, Brunehilde Ezanno, Iris Menage, Melodie Renault, Jan Horst
自由职业者：Katharina Puxbaum
顾问：trans_city - TC, Vienna (A); Christian A. Pichler, Vienna (A); Werner Sobek Ingenieure, Stuttgart (D); Wackler Ingenieure, Birkenfeld (D); TEAMGMI, Vienna (A); Schoberl & Poll, Vienna (A); IMS Brandschutz, Linz (A); Eddie Young, Paris (F); Mattweiss Architekturmodellbau, Vienna (A)

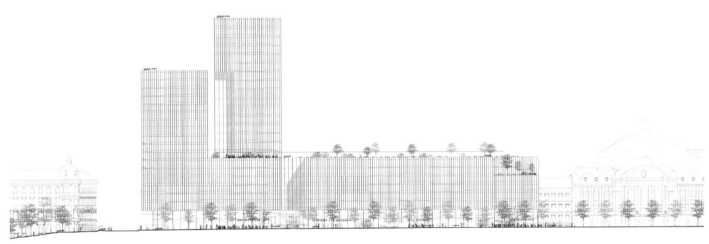
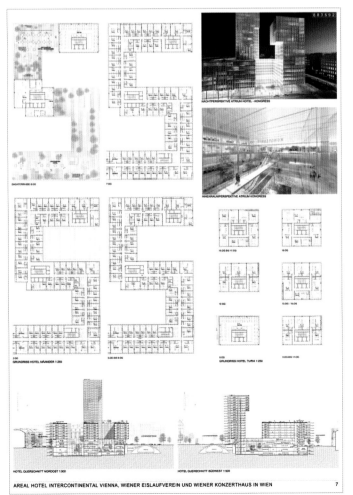
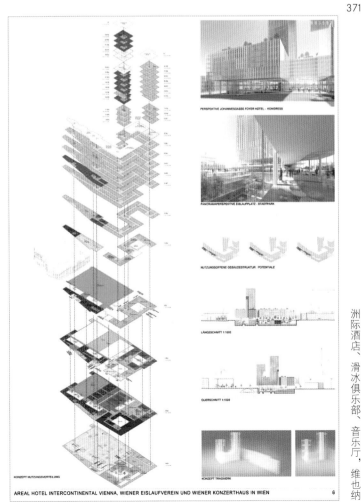

竞赛荣誉奖　获奖事务所

querkraft architekten

Vienna (A)

方案设计者：Jakob Dunkl, Alexandre Mellier
参与者：Johannes Langer, Katarina Jovic, Falk Kremzow, Wanda Gavrilescu
自由职业者：James Diewald, Carmen Hottinger, Dominik Bertl

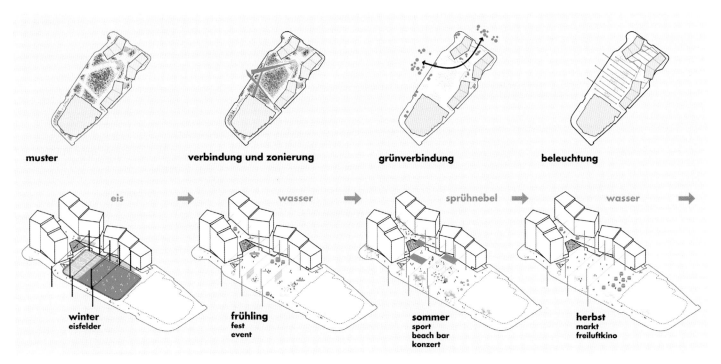

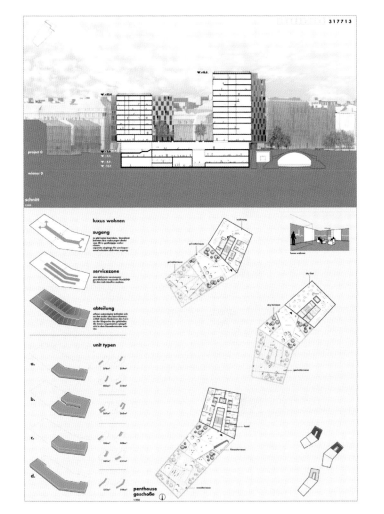

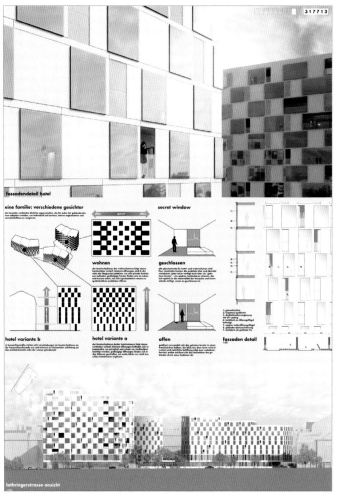

其他竞赛项目

2009—2015 其他竞赛项目的总体回顾

Essen Trade Fair Development
Essen, Germany
2012

Client
MESSE ESSEN GmbH in collaboration
with GVE Grundstücksverwaltung
Stadt Essen GmbH
in cooperation with the city of Essen
1st prize
slapa oberholz pszczulny | architekten
Düsseldorf (D)

Investitionsbank des Landes Brandenburg
Potsdam, Germany
2011-2012

Client
Investitionsbank des Landes Brandenburg
1st prize
KSP Jürgen Engel Architekten
Berlin (D)

Gymnasium+ at Bürgerpark
Lahr, Germany
2014

Client
Stadt Lahr
1st prize
Ackermann + Raff
with Leonhardt, Andrä und Partner
Stuttgart (D)

Facade Competition Staufen-Gallery

Göppingen, Germany
2012–2013

Client
Grundstücksgesellschaft Einkaufszentrum Göppingen mbH, a joint venture company of the ACREST Property Group GmbH and the Familie Schenavsky RS Immobilien Verwaltungs GmbH & Co. KG in cooperation with the city of Göppingen
1st place
von Bock Architekten
Göppingen (D)

New Living in Gartenstadt Falkenberg

Berlin, Germany
2014–2015

Client
Berliner Bau- und Wohnungsgenossenschaft von 1892 eG
1st prize
Anne Lampen Architekten with Dagmar Gast Landschaftsarchitekten
Berlin (D)

Mercator Project

Duisburg, Germany
2010–2011

Client
City of Duisburg
1st prize
Gewers & Pudewill
Berlin (D)

 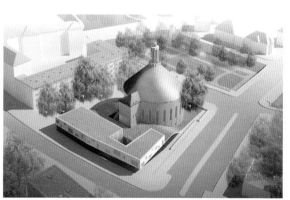

Community Center Paulus Tempelhof

Berlin, Germany
2009–2010

Client
Evangelische Paulus-Kirchengemeinde Tempelhof
1st prize
Kersten + Kopp Architekten
Berlin (D)

附 录

索引

(se)arch – Freie Architekten, Stuttgart (D)	286, 288-289
1010-Architektur, Hanover (D)	260
1X1 Architecture, Winnipeg (CDN)	94-95
3XN A/S, Copenhagen (DK)	51, 186, 190, 334, 340-341, 364
5468796 Architecture, Winnipeg (CDN)	86

A

AAU/A, Paris (F)	66
AB Rozhdestvenka, Moscow (RUS)	334, 344-345
ABD Architects, Moscow (RUS)	334, 346-347
Abdullah Kocamaz, Istanbul (TR)	64
ACD Studio, St. Petersburg (RUS)	67
Acharhabi Architecte, Casablanca (MA)	67
Ackermann + Raff, Stuttgart (D)	374
Adjaye Associates, London (GB)	350
Adler & Olesch Landschaftsarchitekten, Stadtplaner SRL und Ingenieure, Mainz (D)	198-199
AdM Arquitectes, Barcelona (E)	66
Adriano De Gioannis, Rome (I)	66
AECOM, Burnaby/Winnipeg (CDN)	84, 88
Aedas, London (GB)	67
AEP Architekten Eggert Generalplaner, Stuttgart (D)	216
Agence Fikri Benabdallah Architecte, Rabat (MA)	65
Agence TER Payasagistes-Urbanistes, Paris (F)	128-129
Agence Ter.de Landschaftsarchitekten, Karlsruhe (D)	114-115, 364
Agence Up, Paris (F) Zurich	84
agps architecture, Los Angeles (USA)/Zurich (CH)	160, 302, 350
ahrens grabenhorst architekten, Hanover (D)	322
AiB estudi d'arquitectes, Barcelona (E)	66
Aime Kakon, Casablanca (MA)	68, 78-79
AIX Arkitekter, Stockholm (S)	350, 352, 356-357
Akkers Joop Architekt, Wijchen (NL)	294
alexa zahn architekten, Vienna (A)	41, 152
Alexander Paul Architekt, Glienicke (D)	294
all Design, London (GB)	43, 364
AllesWirdGut Architektur, Vienna (A)	160, 162, 178-179, 230
Amstein+Walthert, Karlsruhe (D)/Zurich (CH)	114-115
Andreas J. Keller, Frankfurt am Main (D)	260
andruetto deri architetti associati, Pisa (I)	64
Anna Conti Architetture, Florence (I)	86
Anne Lampen Architekten, Berlin (D)	375
Annette Paul, Potsdam (D)	308
Anton Meyer Architekt, Dachau (D)	153
Arata Isozaki & Associates, Tokyo (J)	350, 352, 354-355
arb east architects, Hanoi (VN)	66
ARCHEA – MALE, Berlin (D)/Florence (I)	322
Architect Eman Hatem, Cairo (ET)	64
Architectural Bureau Zotov & Co, Kiev (UA)	65, 350
Architectural Institute of British Columbia, Vancouver (CDN)	86
architecture & design, Isabelle Gaspard, Paris (F)	67
Architecture & Heritage, Plovdiv (BG)	67
Architecture Project, Valletta (M)	66
Architecture Republic, Dublin (IRL)	66
Architecture Studio V. Schevchenko, Kiev (UA)	350
Architekt Backe, Berlin (D)	260
Architekt Johannes Walther, Hamburg (D)	322
Architekt Peter Schoof, Hanover (D)	260
Architekten Baumgart Wockenfuß und Partner, Celle (D)	260
Architektur- Stadtplanungsbüro, Westerstede (D)	260
Architekturbüro Albert und Planer, Rostock (D)	294
Architekturbüro Albrecht, Munich (D)	65
Architekturbüro Borries + Partner, Rostock (D)	294
Architekturbüro Delugan Meissl, Vienna (A)	11, 300, 304-305, 364
Architekturbüro Lungwitz, Dresden (D)	260
Architekturbüro Ostermeyer, Hanover (D)	258
Architekturbüro Professor Wolfgang Kergaßner, Ostfildern (D)	138, 260
Architekturbüro Törber, Hanover (D)	260
ARCHITEXTURS und Rachid Andaloussi Ben Brahim Office, Rabat (MA)	65
ARCHLAB, Monopoli (I)	65
Arcus Architects, Belfast (GB)	67
AREP Ville, Paris (F)	120, 130-131
argeplan Stadtplaner und Architekten, Hanover (D)	258, 294
ARJM Architecture, Brussels (B)	65
arkiteyp, Istanbul (TR)	65
Arkkitehtitoimisto Lahdelma & Mahlamäki, Helsinki (FIN)	350, 364
Armstrong + Cohen Architecture, Gainesville (USA)	66
Arquitectura y Agronomia, Barcelona (E)	364
Arquivio Architectura, Madrid (E)	64
Arriola & Fiol Arquitectes, Barcelona (E)	66
ARSP, Dornbirn (A)	364
Art de Lux, Munich (D)	294
Arup, International	78-79, 104-109, 112-113, 116-117, 278-279, 280-281, 354-355, 360-361
AS. Architecture Studio, Paris (F)	120, 122-125
AS&P Albert Speer & Partner, Frankfurt am Main (D)	160, 162, 182-183
ASADOV Architectural Studio, Moscow (RUS)	334, 340-341
ASTOC, Cologne (D)	114-115
Asymptote Architecture, New York (USA)	350
ATBA atelier baya, Tiznit (MA)	66
Atelier 30 Architekten, Kassel (D)	51, 206, 210-211
ATELIER 8000, Prague (CZ)	65
Atelier Anonymous, Vancouver (CDN)	86
Atelier d'architecture Chaix & Morel et Associés, Paris (F)	10, 11, 45, 138, 140, 144-145, 364, 366, 370-371
Atelier Kempe Thill, Rotterdam (NL)	160, 162, 174-177
Atelier Loidl Landschaftsarchitekten, Berlin (D)	252-253, 364, 366
Atelier Moto Katono, Tokyo (J)	65
Atelier Paolo Piva, Vienna (A)	364
Atelier ST Architekten, Leipzig (D)	231
Atelier Thomas Pucher, Graz (A)	153
Atelier3AM, Toronto (CDN)	68, 80-81
ATP Architekten und Ingenieure, Innsbruck (A)	216
Auböck + Kárász Landschaftsarchitekten, Vienna (A)	364
Auer + Weber + Assoziierte, Munich/Stuttgart (D)	120, 128-129, 138, 188
augustin und frank architekten, Berlin (D)	300
Aziza Chaouni Projects, Toronto (CDN)	86

B

B+H Architects, Toronto (CDN) — 86
Bahia Nouh, Fez (MA) — 68, 76–77
Bahl + Partner Architekten, Hagen (D) — 258, 262
Bär Stadelmann Stöcker Architekten, Nuremberg (D) — 10, 11
Barkow Leibinger Architekten, Berlin (D) — 138
Bassen Sauletshileri, Almaty (KZ) — 66
bauchplan).(, Munich (D) — 364
Baumschlager Eberle Gruppe | BE Berlin, Berlin (D)/ Lochau (A) — 188, 206, 216, 230, 242
BBCF Architectes, Paris (F) — 294
bbz Landschaftsarchitekten Berlin, Berlin (D) — 230
Behles & Jochimsen Architekten, Berlin (D) — 11, 258, 262
Behnisch Architekten, Stuttgart (D) — 47, 196, 206, 266, 270–273
Berger + Parkkinen Architekten, Vienna (A) — 65, 364
bgf+ architekten, Wiesbaden (D) — 186, 190
BHBVT Architekten, Berlin (D) — 206, 230
BIG – Bjarke Ingels Group, Copenhagen (DK) — 160, 188
blauraum architekten, Hamburg (D) — 47, 240
BLK2 Böge Lindner K2 Architekten, Hamburg (D) — 230
Blocher Blocher Partners, Stuttgart (D) — 47, 268
Blumers Architekten Generalplanung und Baumanagement, Berlin (D) — 248
Bodamer Faber Architekten, Stuttgart (D) — 230, 236–237
Böge Lindner Architekten, Hamburg (D) — 188, 258, 262
Bothe Richter Teherani Architekten, Hamburg (D) — 268
Boyarsky Murphy Architects, London (GB) — 64
Brechensbauer Weinhart + Partner Architekten, Munich (D) — 216
Breimann & Bruun Landschaftsarchitekten, Hamburg (D) — 84, 312, 318–319
brh Architekten + Ingenieure, Berlin (D) — 248
Brown and Storey Architects, Toronto (CDN) — 86
Brullet - De Luna Arquitectes, Barcelona (E) — 206
Brunet Saunier Architecture, Paris (F) — 130–131, 216, 218, 226–227
Brunhart Brunner Kranz Architekten, Balzers (FL) — 322
Bruno Fioretti Marquez Architekten, Berlin (D) — 11, 47, 230, 240, 244–245
Brut Deluxe, Madrid (E) — 67
Buchner Bründler Architekten, Basel (CH) — 11
Bularch, Sofia (B) — 64
Bureau E.A.S.T., Los Angeles (USA)/Fez (MA) — 68, 80–81
Burger Landschaftsarchitekten Partnerschaft, Munich (D) — 364
Burgos & Garrido Arquitectos Asociados, Madrid (E) — 84
Buro Bol, 's-Hertogenbosch (NL) — 84
Büro Luchterhandt Stadtplanung, Hamburg (D) — 294
BVN Donovan Hill, Melbourne (AUS) — 86
BWM Architekten und Partner, Vienna (A) — 364

C

C+S Architects, Treviso (I) — 84, 364
Cabinet Hadmi, Safi (MA) — 68
cabinet Seqqat Nabila, Meknès (MA) — 64
CAMPO aud, Rio de Janeiro (BR) — 66
Campus 2 Cover, Moralzarzal (E) — 86
capatti staubach Urbane Landschaften, Berlin (D) — 230
caramel architekten, Vienna (A) — 86
Caravaggi Lucina Architect, Rome (I) — 134–135
Carlos Arroyo Architects, Madrid (E) — 67
Carmelo Bagalà Architects, Milan (I) — 66
CASALEGANITOS Estudio de Arquitectura, Madrid (E) — 67
CAVstudio, Lisbon (P) — 66
Christoph Faulhaber, Hamburg (D) — 308
Cibinel Architecture, Winnipeg (CDN) — 90–93
CIVITAS Urban Design and Planning, Vancouver (CDN) — 86
CKM, Singapore (SGP) — 66
Claudia Weber-Molenaar, Gräfelfing (D) — 230
Claudio Silvestrin Architects, London (GP) — 65
claudiovilarinho.com architects and designers, Porto (P) — 66
Claus en Kaan Architecten, Rotterdam (NL) — 138, 350
Clive Wilkinson Architects, Culver City (USA) — 41, 160
club L94 Landschaftsarchitekten, Cologne (D) — 148–149, 160, 162, 178–179, 285
CO A. Architektenkooperative, Berlin — 322, 328–329
CODE UNIQUE, Dresden (D) — 230, 232–233
Coelacanth and Associates (C+A), Nagoya (J) — 120, 126–127
Cohlmeyer Architecture, Winnipeg (CDN) — 96–97

Consolidated Consultants - Jafar Tukan Architects, Amman (JOR) — 66
ContisNambiar, Brooklyn (USA) — 86
Coop Himmelb(l)au, Vienna (A) — 10, 41, 364
Cornelsen + Seelinger Architekten, Darmstadt (D) — 258, 262
CPG Consultants, Singapore (SGP) — 120, 132–133
Creative Landscapes, Naples (I) — 86
Czerner Göttsch Architekten, Hamburg (D) — 294

D

D'Appolonia, Genoa (I) — 120, 134–135
Dagmar Gast Landschaftsarchitekten, Berlin (D) — 375
David Chipperfield Architects, Berlin (D) — 10, 11, 350, 352, 358–359
Davide Rapp, Milan (I) — 65
DE FABRIEK, Nijmegen (NL) — 294
DE+ Architekten, Berlin (D) — 292, 296
Degelo Architekten, Basel (CH) — 138
dennis ulm architekt, Munich (D) — 65
Deubzer König Architekten, Berlin (D) — 294
diaspora, Coimbra (P) — 66
Didrihsons un Didrihsons, Riga (LV) — 86
Dietrich | Untertrifaller Architekten, Bregenz/Vienna (A) — 153, 350, 352
Dinmez Insaat Sanayi, Izmir (TR) — 66
disart, Barcelona (E) — 67
DnD Landschaftsplanung, Vienna (A) — 364, 366, 370–371
Dolenc Scheiwiller Architekten, Zurich (CH) — 302
Dominique Perrault Architecture, Paris (F) — 350
doranth post architekten, Munich (D) — 196
DP Architects, Singapore (SGP) — 66, 160
Dr. Heinekamp Labor- und Institutsplanung, Berlin (D) — 200–201
Dr. Maged Aboul-Ela & Architect Eman Hatem, Cairo (ET) — 64
DRC, Jounieh (RL) — 66
Drexler Guinand Jausin, Frankfurt am Main (D) — 260
DRMS, Amsterdam (NL) — 66
DTAH, Toronto (CDN) — 84, 88, 96–97
DUDA/PAINE Architects, Durham (USA) — 86
Durand-Hollis Rupe Architects, San Antonio (USA) — 65
DURU & PARTNERS, Montpellier (F) — 65

E

EBA [For Cosmopolitain], Toronto (CDN) — 66
ECOTONelu Design Studio, Seongnam (ROK) — 86
el:ch landschaftsarchitekten, Berlin (D) — 156–157, 312
Eleena Jamil Architect, Ampang (MAL) — 66
elementa architects, Seoul (ROK) — 66
Eller + Eller Architekten, Düsseldorf (D) — 186, 190, 192–193, 268, 282–283
els architecture, Decatur (USA) — 67
Emiliano Bugatti, Istanbul (TR) — 68
energydesign, Shanghai (CN) — 126–127
Enlace Arquitectura, Caracas (YV) — 65
ernst niklaus fausch architekten, Zurich (CH) — 160, 162, 170–173
Estudi d'Arquitectura Josep Blesa, Valencia (E) — 66
Estudio CV_Public Opera, Buenos Aires (RA) — 66
Eurich Gula Landschaftsarchitekten, Wendlingen (D) — 200–201
EWA, London (GB) — 66

F

fabriK·B Architekten, Berlin (D) — 294
FACE2050, Bad Griesbach (D) — 66
FARSHID MOUSSAVI ARCHITECTURE, London (GB) — 43, 334, 344–345
Federico Wulff & Melina Guirnaldos Arquitectos, Madrid (E) — 65
Ferdinand Heide Architekt, Frankfurt am Main (D) — 11, 188
FGA Arquitectos, Mexico Stadt (MEX) — 64
Filipe Oliveira Dias Arquitecto, Porto (P) — 65
Fink + Jocher Architekten und Stadtplaner, Munich (D) — 242
Florian Dombois, Cologne (D) — 308
Forsthuber & Martinek Architekten, Salzburg (A) — 286
Foster and Partners, London (GB) — 268
Francesca Mugnai Architetto, Florence (I) — 66
Franck - Von Reusner, Lübeck (D) — 294
Frank Architekten, Munich (D) — 294
Frank Görge Architekt, Hamburg (D) — 294
Frank Kiessling Landschaftsarchitekten, Berlin (D) — 364
Franke Seiffert, Stuttgart (D) — 286
Freie Architektin Jeong, Stuttgart (D) — 294

Freimüller Söllinger Architektur, Vienna (A)	364
Fritsch + Tschaidse Architekten, Munich (D)	196
Fritz-Dieter Tollé Architekten, Verden (D)	258
Fritzen + Müller-Giebeler Architekten, Ahlen (D)	258, 262
FSA architects, Madrid (E)	66
FSWLA, Düsseldorf (D)	153
FT3 Architecture Landscape Interior Design, Winnipeg (CDN)	86
Fugmann Janotta Landschaftsarchitektur und Landschaftsentwicklung, Berlin (D)	230
Fundacion CEPA, Buenos Aires (RA)	66
Füsun Türetken, Berlin (D)	308

G

Gaheez Consultants, Peshawar (PK)	67
Gantert + Wiemeler Ingenieurplanung, Münster (D)	200-201
Gasparini Christian Architekt, Reggio Emilia (I)	294
Gatermann + Schossig Bauplanungsgesellschaft, Cologne (D)	138, 188, 230
Gauselmann Soll Architekten, Dortmund (D)	294
Gerald Hofmann, Nuremberg (D)	308
Gerber Architekten, Dortmund (D)	11, 188
Gesa Königstein Landschaftsarchitektin, Berlin (D)	294
Geske-Wenzel Architekten, Berlin (D)	294
geskes.hack Landschaftsarchitekten, Berlin (D)	312
Geurst & Schulze Architecten, Den Haag (D)	64
Gewers & Pudewill, Berlin (D)	375
GFK Gesellschaft für Krankenhausberatung, Cologne (D)	226-227
gh3, Toronto (CDN)	312
GHP Landschaftsarchitekten, Hamburg (D)	231
Glass Kramer Löbbert Architekten, Berlin (D)	206
GLP Ingenieurgesellschaft, Hamburg (D)	200-201
gmp, Aachen/Berlin/Frankfurt am Main/Hamburg (D)	84, 138, 160, 162, 180-181, 186, 190, 196, 206, 216, 218, 230, 234-235, 268
GMS Freie Architekten, Isny im Allgäu (D)	260
Gnadler.Meyn.Woitassek Architekten Innenarchitekten, Stralsund (D)	292, 296
Goetz Hootz Castorph Planungsgesellschaft, Munich (D)	153
Götz Lemberg, Berlin (D)	308
Graeme Massie Architects, Edinburgh (GB)	66
Graft Architekten, Berlin (D)	302
Grigat-Architekten, Stadthagen (D)	260
Groupe3Architectes, Rabat (MA)	65
Grüntuch Ernst Architekten, Berlin (D)	138, 140, 146-147
Grüttner Architekten, Soest (D)	260
guillaume girod architecte, Grenoble (F)	66
Guillermo Vázquez Consuegra Arquitecto, Sevilla (E)	364

H

h.s.d. architekten, Lemgo (D)	153
H3T Architekti, Prague (CZ)	65
h4a Gessert + Randecker + Legner Architekten, Stuttgart (D)	230
HAAS Architekten, Berlin (D)	47, 240, 248
Hadi Teherani Architects, Hamburg (D)	153
Hager Partner, Berlin (D)	231, 250-251
hammeskrause architekten, Stuttgart (D)	45, 49, 196, 200-201
HANDIS, Casablanca (MA)	66
Hanke + Partner Landschaftsarchitekten, Berlin (D)	231
Hanoi Design Construction, Hanoi (VN)	132-133
Hans-Hermann Krafft, Berlin (D)	364
hanse unit, Hamburg (D)	64, 68
Happold Ingenieurbüro, Berlin (D) / London (GB)	104-109, 156-157
Haptic, London (GB)	65
harris + kurrle architekten, Stuttgart (D)	286
Harter + Kanzler Freie Architekten, Freiburg (D)	286
Hascher Jehle, Berlin (D)	49, 138, 141, 206, 212-213, 216, 231
Hashim Sarkis Studios, Beirut (RL)	64, 68, 80-81
HBS, Hanoi (VN)	128-129
HCP Architecture & Engineering, Malaga (E)	66
HDAA – Heitor Derbli Arquitetos Associados, Rio de Janeiro (BR)	67
HDR TMK Planungsgesellschaft, Düsseldorf (D)	216
Heide & von Beckerath Architekten, Berlin (D)	248
Heidenreich & Springer Architekten, Berlin (D)	248, 254-255
Heine Mildner Architekten, Dresden (D)	292, 296
Heinisch Landschaftsarchitekten, Gotha (D)	230
Heinke Haberland, Düsseldorf (D)	308
Heinle, Wischer und Partner, Berlin/Cologne (D)	45, 196, 216, 218, 220-221, 231
heneghan.peng architects, Dublin (IRL)	322
Henke Schreieck Architekten, Vienna (A)	364
Henn Architekten, Munich (D)	45, 53, 102, 116-117, 160, 162, 164-169, 196
Henning Larsen Architects, Copenhagen (DK)/Munich (D)	45, 152, 156-157, 196, 268
Hernandez Leon Arquitectos, Madrid (E)	67
herrburg Landschaftsarchitekten, Berlin (D)	230, 232-233
Hester Oerlemans, Berlin (D)	308
HG Merz, Berlin (D)	322
hildebrandt.lay.architekten, Berlin (D)	242
Hilderman Thomas Frank Cram Landscape Architecture & Planning, Winnipeg (CDN)	86
hks Hestermann Rommel Architekten, Erfurt (D)	230
HL-PP Ingenieure International, Munich (D)	128-129
HOK International, Hong Kong (HK)	64
HOSHINO ARCHITECTS, Tokyo (J)	66
HSP Hoppe Sommer Planungsgesellschaft, Stuttgart (D)	216
Hübotter + Stürken Architektengemeinschaft, Hanover (D)	292, 296
HWP Planungsgesellschaft, Stuttgart (D)	216, 218

I

IAD Independent Architectural Diplomacy, Madrid (E)	51, 84, 88, 98-99
IBB Ingenieurbüro, Leipzig (D)	202-203
icarquitectura, Figueres (E)	67
Idealice – technisches büro für landschaftsarchitektur, Vienna (A)	230
IDEAS (ESTUDIO), Rome (I)	134-135
IDEAS, Milan (I)	134-135
Idrissi Architecture Office, Salé (MA)	65
IF architecture, Hanover (D)	260
IGF Ingenieurgesellschaft Feldmeier, Münster (D)	200-201
Inbo, Amsterdam (NL)	84
Independents 499, Buffalo (USA)	66
Ingenieurbüro Dr. Binnewies Ingenieurgesellschaft, Hamburg (D)	114-115
Ingenieurbüro Horn + Horn, Neumünster (D)	222-223
INGEROP International, Courbevoie (F)	122-123
INROS LACKNER, Rostock (D)	294
Integral Group, Toronto (CDN)	96-97
iproplan Planungsgesellschaft, Ho Chi Minh City (VN)	126-127
Isay Weinfeld Arquitectos e Urbanismo, São Paulo (BR)	364, 366, 368-369
ISW – Ingenieur Schmidt & Willmes, Hamm (D)	226-227
Itsuko Hasegawa Atelier, Tokyo (J)	102, 112-113
Itten+Brechbühl, Berlin (D)	206
Ivan Kroupa Architekti, Prague (CZ)	350
IWP Ingenieurbüro für Systemplanung, Stuttgart (D)	198-199
IXII, Kobe (J)	64

J

J. Deutler/Architekturbüro R. Schacht, Rostock (D)	294
Jabusch + Schneider Architekten, Hanover (D)	258, 262
Jakob Timpe, Berlin (D)	258
Janet Rosenberg & Studio, Toronto (CDN)	84, 88, 90-93
Janin Rabaschus Architektin, Dresden (E)	294
JCSR Arquitecto, Sevilla (E)	65
Jean-Yves QUAY, Lyon (F)	66
Jetter Landschaftsarchitekten, Stuttgart (D)	231
John McAslan + Partners, London (GB)	334, 338-339
Josep Sánchez Ferré Architekt, Barcelona (E)	294
JSWD Architekten, Cologne (D)	10, 11, 47, 138, 140, 144-145
Jürgen Scharlach Architektur und Stadtplanung, Isernhagen (D)	258
JURONG Consultants, Singapore (SGP)	66
JZMK PARTNERS, Irvine (USA)	64

K

K.N.Z design architecture & space, Amman (JOR)	66
KAAN Architekten, Munich (D)	47, 160
kadawittfeldarchitektur, Aachen (D)	138, 188
Kai Lorberg Architekt, Hamburg (D)	294
Karres en Brands Landschapsarchitecten, Hilversum (NL)	312, 314-315
Karsten K. Krebs Architekten, Hanover (D)	258, 262
Kaspar Kraemer Architekten, Cologne (D)	41, 138, 140, 148-149
KE architekten und Francesco Minnitti, Winterthur (CH)	65

Keith Williams Architects, London (GB) 84
Kemper Steiner & Partner Architekten + Stadtplaner, Bochum (D) 216, 218, 226–227
Kengo Kuma & Associates, Tokyo (J) 66
Kersten + Kopp Architekten, Berlin (D) 375
Kirk + Specht Landschaftsarchitekten, Berlin (D) 292, 296
Klaus Kada, Graz (A) 350
Kleihues + Kleihues Architekten, Berlin (D) 11, 153, 266, 274–275, 350, 352, 360–361
Klein und Sänger Architekten, Munich (D) 230
Klingmann Architects, New York (USA) 66
KOKO architects, Tallinn (EST) 65
kolb hader architekten, Vienna (A) 64, 68
KOLLIAS GEORGE, Heraklion (GR) 65
Kölling Architekten, Bad Vilbel (D) 260
Köppler Türk Architekten, Berlin (D) 294
Koschany Zimmer Architekten KZA, Essen (D) 66
Kraaijvanger · Urbis, Rotterdam (NL) 64, 322
Krätzig & Partner Ingenieurgesellschaft für Bautechnik, Bochum (D) 226–227
KSP Jürgen Engel Architekten, Berlin/Frankfurt am Main (D) 53, 102, 110–111, 152, 374
KSV Krüger Schuberth Vandreike, Berlin (D) 231
Kubik Studio, Meknes (MA) 68
Kuehn Malvezzi, Berlin (D) 64, 188
KuKu, Athens (GR) 66
Kummer.Lubk.Partner, Erfurt (D) 230
Kunzemann Architekten, Großburgwedel (D) 258
Kusus + Kusus Architekten, Berlin (D) 138

L

L'OEUF, Montreal (CDN) 86
L/A Liebel/Architekten, Aalen (D) 286
LA.BAR Landschaftsarchitekten, Berlin (D) 230
LAAP Landscape + Architecture, Arquitectura + Paisaje, Mexiko-Stadt (MEX) 65
LAND IN SICHT – Büro für Landschaftsplanung, Vienna (A) 364, 366
Landmark Planning & Design, Winnipeg (CDN) 90–93
Langdon & Seah Vietnam, Hanoi (VN) 126–127, 128–129
Larkin Architect, Toronto (CDN) 64
Latz + Partner LandschaftsArchitekten Stadtplaner, Kranzberg (D) 224–225, 364
lauth : van holst architekten, Wiesbaden (D) 322
lbgo architektur, Munich (D) 286
LDA Design, London (GB) 364
LEESER Architecture, Brooklyn (USA) 334, 346–347
Lehrecke Architekten, Berlin (D) 11
Léon Wohlhage Wernik Architekten, Berlin (D) 11, 242
Leonhardt, Andrä und Partner, Stuttgart (D) 374
Levin Monsigny Landschaftsarchitekten, Berlin (D) 104–109, 202–203
Lieseberg Architekten, Hanover (D) 258
Limin Hee, Singapore (SGP) 66
Lino Bianco and Associates, Hamrun (M) 66
Lips + Teichert Architekten, Freiburg-March (D) 292, 296
Lisa Kimling, Freiburg (D) 286
LOADINGDOCK5 ARCHITECTURE, New York (USA) 65
Locke Lührs Architektinnen, Dresden (D) 294
LOHANATA Design, Jakarta (RI) 86
Lohaus + Carl Landschaftsarchitekten, Hanover (D) 292
Lostmodern, Paris (F) 66
Lothar Jeromin Architekt, Essen (D) 153
LOVE architecture and urbanism, Graz (A) 11, 41, 49, 66, 152, 154–155
lüderwaldt architekten, Hanover (D) 260
LUDES Architekten - Ingenieure, Recklinghausen (D) 216
Ludes Generalplaner, Berlin (D) 216, 218, 222–223

M

ma.lo architectural office, Innsbruck (A) 65
Machado and Silvetti Associates, Boston (USA) 102, 104–109, 364
Magma Architecture, Berlin (D) 258
Mahmoud Saimeh, Amman (JOR) 67
male architekten, Berlin (D) 258
Man Made Land, Berlin (D) 230, 234–235
Manfredi Anello, Dublin (IRL) 258

manzl ritsch sandner architekten, Innsbruck (A) 66
Marc Bausback, Berlin (D) 308
Marcel Adam Landschaftsarchitekten, Potsdam (D) 230
Marek Jahnke Landschaftsarchitekt, Berlin (D) 231, 292, 296
mari-as arquitectos asociados, Sevilla (E) 64
Marid Terzic, Vienna (A) 364
Marije Tersteege, Amsterdam (NL) 66
Markus Fiegl Architekt, Berlin (D) 292, 296
Markus Klink, Stuttgart (D) 308
Marte.Marte Architekten, Weiler (D) 322, 326–327
Martienssen Architekten, Hanover (D) 258
MASSIMILIANO FUKSAS Architetto, Rome (I) 49, 334, 336–337
Mathes Beratende Ingenieure, Chemnitz (D) 224–225
matrix architektur, Rostock (D) 292, 296
Matthias Lanzendorf Landschaftsarchitekten, Leipzig (D) 231
matzke | architekten, Berlin (D) 294
Max Dudler Architekten, Berlin (D) 138, 153, 188, 322, 324–325, 364, 366
maxwan, Rotterdam (NL) 64
Mayr|Ludescher|Partner Beratende Ingenieure, Stuttgart (D) 198–199
MECANOO International, Delft (NL) 51, 334, 342–343
meck architekten, Prof. A. Meck Architekturbüro, Munich (D) 138
Meier-Scupin + Partner Architekten, Munich (D) 260
MEMA Arquitectos, Bogota (CO) 66
Meyer-Wolters & Yeger Architekten, Hamburg (D) 258, 262
Mezger & Schleicher, Stuttgart (D) 65
MGF Architekten, Stuttgart (D) 188
mhb Planungs- und Ingenieurgesellschaft, Rostock (D) 292, 296
Mijic Architects, Rimini (I) 258, 262
Mila/Jakob Tigges, Berlin (D) 65, 84, 138
mm architekten, Hanover (D) 51, 258, 262
MM26, Padua (I) 64
MMM Group Limited, Ottawa (CDN) 86
MMZ Architekten, Hanover (D) 258
MOBA Studio, Prague (CZ) 66
Modul PKB, St. Petersburg (RUS) 84
moh architects, Vienna (A) 66
Mola + Winkelmüller Architekten, Berlin (D) 242, 258
Molestina Architekten, Cologne (D) 153
Monnerjan · Kast · Walter Architekten, Düsseldorf (D) 216
mossessian & partners architecture, London (GB) 11, 64, 68, 70–75
Moxon Architects, London (GB) 64, 68, 78–79, 84
MRSCHMIDT Architekten, Berlin (D) 294
msm meyer schmitz-morkramer, Darmstadt (D) 186, 190
Müller Illien Landschaftsarchitekten, Zurich (CH) 160, 162, 170–173
Müller-Born-Architekten, Kassel (D) 294
Museum The Garage, Rotterdam (NL) 65

N

N.E.E.D., New York (USA) 65
N.I.L. Ingenieurgesellschaft, Berlin (D) 222–223
N+B architectes, Elodie Nourrigat & Jacques Brion, Montpellier (F) 64
National General Construction Consulting, Ho Chi Minh City (VN) 130–131
NED University of Engineering & Technology, Karachi (PK) 66
Neutelings Riedijk Architecten, Rotterdam (NL) 364
Never Ending Architecture, N.E.A., Jerusalem (IL) 64
Nickl & Partner Architekten, Munich (D) 196, 216, 218, 224–225, 231, 260
Nieberg Architekten, Hanover (D) 260
Nieto Sobejano Arquitectos, Berlin (D)/Madrid (E) 65, 138, 186, 190, 322, 364
NO.MAD Arquitectos, Madrid (E) 153
NOA Architecture, New York (USA) 86
nodo17 Architects, Madrid (E) 84, 88
Nopto Architekt, Herzebrock-Clarholz (D) 294
nps tchoban voss, Berlin (D) 258, 262
Number TEN Architectural Group, Winnipeg (CDN) 86
Numrich Albrecht Klumpp Architekten, Berlin (D) 230

O

OD205SL, Delft (NL) 84
OKRA Landschapsarchitecten, Utrecht (NL) 160
Olekov Architects, Sofia (BG) 64
OMAYAN, Tangier (MA) 65
ORG-Design & Architecture, Hanover (D) 260
ORTIZ MONASTERIO + ASOCIADO, Mexico City (MEX) 64

Ortner & Ortner Baukunst, Berlin (D)/Vienna (A)	138, 364, 366
OSD, Frankfurt am Main (D)	202-203
Oskar Leo Kaufmann | Albert Rüf, Dornbirn (A)	138

P

P.arc, Berlin (D)	206
P.I.A – Architekten, Karlsruhe (D)	286
P.T. Morimura & Associates, Tokyo (J)	112-113
Pannett & Locher Architekten, Bern (CH)	294
Papazian Roy Architecte, Paris (F)	67
Pascal Flammer Büro für Architektur, Zurich (CH)	66
Pascal's Limaçon Creative Teamwork, St. Petersburg (RUS)	84
Paul Bretz Architekten, Luxemburg (L)	260
paula santos, arquitectura, Porto (P)	66
Paulo David Arquitecto, Madeira (P)	350
pbr Planungsbüro Rohling Architekten und Ingenieure, Braunschweig/Osnabrück (D)	64, 216
Perkins+Will, Vancouver (CDN)	53, 84, 88, 94-95
Peter Kellow Architecture, Plymouth (GB)	65
Peter Kulka Architektur, Dresden/Cologne (D)	138, 307, 322
Peter Sandhaus, Berlin (D)	308
Peter Tagiuri, Architects, Cambridge (GB)	65
PFS Studio, Vancouver (CDN)	94-95
Pich-Aguilera Arquitectos, Barcelona (E)	206
Pinearq, Barcelona (E)	206
PLACEMEDIA, Landscape Architects Collaborative, Tokyo (J)	126-127
Plain Projects, Winnipeg (CDN)	86
plandrei Landschaftsarchitektur, Erfurt (D)	230
Plankontor S1 Landschaftsarchitekten, Stuttgart (D)	230, 236-237
Planstatt Senner, Überlingen (D)	230
POLY RYTHMIC ARCHITECTURE, Bordeaux (F)	66
Poos Isensee Architekten, Hanover (D)	258
Prof. Kollhoff Generalplanung, Berlin (D)	260
Project BASE, Paris (F)	122-123
Project Meganom, Moscow (RUS)	334, 338-339
PurserLee, Dallas (USA)	64
PZP ARHITECTURA, Bucharest (RO)	66

Q

querkraft architekten, Vienna (A)	41, 364, 366, 372-373

R

R&Sie(n) Architects, Paris (F)	65
r10r10 Architects, Stuttgart (D)	294
Rafael Viñoly Architects, New York (USA)	45, 268, 282-283
Rafal Mroczkowski Architekci, Poznan (PL)	67
Rainer Schmidt Landschaftsarchitekten, Berlin (D)	222-223
Rapp Infra, Basel (CH)	160, 162, 170-173
Rapp+Rapp, Berlin (D)	188
Rarcon, Vila Nova de Gaia (P)	66
Reaction Architecture, Tunis (TN)	65
Reardon Smith Architects, London (GB)	64
Relative Form Architecture Studio, Vancouver (CDN)	67
Rentschler und Riedesser Ingenieurgesellschaft, Berlin (D)	220-221
Reset architecture, 's-Hertogenbosch (NL)	66, 84
REZERVE, Moscow (RUS)	334, 342-343
Richard Burck Associates, Somerville (USA)	364
Richard Meier & Partner, New York (USA)	268, 280-281
Riegler Riewe Architekten, Graz (A)	102, 114-115, 206
Ring Architekten, Munich (D)	258, 262
Ritzen Architecten, Maastricht (NL)	65
RMP Stephan Lenzen Landschaftsarchitekten, Bonn (D)	220-221
ROBERTNEUN™ Architekten, Berlin (D)	39, 248, 252-253
Rohdecan Architekten, Dresden (D)	206, 208-209
Roland Unterbusch Architekt, Rostock (D)	294
RSE Landscape Architecture, Amsterdam (NL)	86
Rudy Uytenhaak Architectenbureau, Amsterdam (NL)	196
Runge Architekten, Hanover (D)	260

S

S.A.E.C., Naples (I)	65
Sadar Vuga Arhitekti, Ljubljana (SLO)	350
SAM Architects and Partners, Zurich (CH)	84
SARMA & NORDE Architects, Riga (LV)	65
Sauerbruch Hutton Architekten, Berlin (D)	153, 266, 276-277, 350, 364
SB-Studio, Saarbrücken (D)	260
SBarch Architetti Associati, Rome (I)	84
scapelab, Ljubljana (SLO)	64
Scharf und Wolf, Berlin (D)	294
Schlaich Bergermann und Partner, Stuttgart (D)	126-127, 146-147
Schlosser | Schlosser, Berlin (D)	292, 296
Schmidt Hammer Lassen Architects, Copenhagen (DK)	350, 352
Schneider + Sendelbach Architekten, Braunschweig (D)	64
Schoppe + Partner Freiraumplanung, Hamburg (D)	230
Schultes Frank Architekten, Berlin (D)	33, 34, 56, 138, 141, 294
Schuster Pechtold Schmidt Architekten, Munich (D)	216
Schweger, Hamburg (D)	138, 140, 142-143, 188, 350, 352
Seo-Kang Architects Office, Seoul (ROK)	67
Serero Architectes Urbanistes, Paris (F)	66
Sergio Pascolo Architects, Venice (I)	65, 260
Shahla Shahmoradi, Teheran (IR)	86
SHArchs, Cincinnati (USA)	66
SIC Arquitectura y Urbanismo, Madrid (E)	67
Sirius Lighting Office, Tokyo (J)	112-113
slapa oberholz pszczulny | architekten, Düsseldorf (D)	374
SMAQ – architecture urbanism research, Berlin (D)	292, 296
SMC Management Contractors, Nicola Fazio, Winterthur (CH)	65
Snøhetta, Oslo (N)	35, 364
SOLIDUM, Medellin (CO)	67
SOW Planungsgruppe, Berlin (D)	196, 202-203
SPEECH, Moscow (RUS)	334, 336-337
spine architects, Hamburg (D)	258
ST raum a. Landschaftsarchitekten, Berlin (D)	160, 162, 180-181, 192-193, 364
Staab Architekten, Berlin (D)	138, 300, 322, 330-331
Standler/Zimmermann Landschaftsarchitektur, Vienna (A)	364
Stankovic Architekten, Berlin (D)	114-115, 126-127, 258, 262
Steidle Architekten, Munich (D)	138
Steiner Weißenberger Architekten, Berlin (D)	292, 296
Stephan Braunfels Architekten, Berlin (D)	138, 258, 262, 322
Stephen Collier Architects, Surry Hills (GB)	67
Steven Fong, Toronto (CDN)	84
stocker dewes architekten, Freiburg (D)	286
Stoss Landscape Urbanism, Boston (USA)	160
STOY – Architekten, Neumünster (D)	292, 296
Studio Ferretti-Marcelloni, Rome (I)	64, 68, 76-77
Studio Giorgio Ciarallo, Rho (I)	64, 68
Studio K, Naples (I)	66
Studio Milou Architecture, Paris (F)	132-133
Studyo Architects, Cologne (D)	260
Sudarch, Reggio Calabria (I)	67
Susanne Dieckmann Architektin, Weimar (D)	294
Süss Beratende Ingenieure, Nuremberg (D)	224-225
swap architekten, Vienna (A)	152
SYMplan Landschaftsarchitekturbüro, Essen (D)	226-227

T

T.A. Wolf Architekten, Munich (D)	294
t17 Landschaftsarchitekten, Munich (D)	160, 230
Taller 301, Bogota (CO)	64
Taoufik El Oufir Architectes, Rabat (MA)	68
Taylor Cullity Lethlean, Victoria (AUS)	112-113
TD, Flachau (A)	86
Team Li Sa, Hanover (D)	260
TENDANCES, Tunis (TN)	65
The Commons, Montreal (CDN)	86
the fourth dimension, Zarinshahr (IR)	67
thoma architekten, Berlin (D)	230
Thomas Eller, Berlin (D)	308
Thomas Möller Architekt, Karlsruhe (D)	294
Thomas Müller Ivan Reimann Architekten, Berlin (D)	138, 141, 152, 230, 248
Thus Ton Städtebauer & Landschaftsarchitekt, Ubbergen (NL)	294
Tokyo Institute of Technology, Hamamatsu (J)	112-113
tönies+schroeter+jansen freie architekten, Lübeck (D)	216
Topotek 1 Landschaftsarchitekten, Berlin (D)	116-117, 160, 162, 164-169
Transsolar Energietechnik, Munich (D)	80-81, 114-115, 178-179, 182-183, 234-235, 270-273

Transver, Munich (D)	162, 174-177
TREUSCH architecture, Vienna (A)	230
TRYS A.M. Architects, Vilnius (LT)	86
TSARA Architectes, Clichy (F)	65
TURATH, Amman (JOR)	66
Turkali Architekten, Frankfurt am Main (D)	188

U

UBIK Architects, Hanoi (VN)	128-129
Ünsal Demir, Istanbul (TR)	66
UNStudio, Amsterdam (NL)	39, 268, 278-279
URBAMED, Paris (F)	66
Urban Edge Consultants, Rahway (USA)	66
Uwe Becker, Berlin (D)	294
Uwe Bernd Friedemann, Cologne (D)	66

V

Vaknine Architects & Town Planners, Jerusalem (IL)	65
VALERIO MORABITO, Reggio Calabria (I)	66
ver.de Landschaftsarchitektur, Freising (D)	160, 162, 182-183
VHA Architects, Hanoi (VN)	122-123
Vogt Landschaftsarchitekten, Berlin (D)/Zurich (CH)	160, 230, 244-245
Volkmar Nickol, Berlin (D)	294
von Bock Architekten, Göppingen (D)	375

W

Walter Gebhardt Architekt, Hamburg (D)	258, 262
Walter Huber Architekten, Stuttgart (D)	286
WAP Architects, Sheffield (GB)	66
WBP Ingenieure für Haustechnik, Münster (D)	202-203
Weber Hofer Partner Architekten, Zurich (CH)	188
Weidinger Landschaftsarchitekten, Berlin (D)	231, 312, 316-317
Weinmiller Architekten, Berlin (D)	230
Werner Sobek, London (GB)	142-143, 144-145, 268, 304-305, 370-371
WES LandschaftsArchitektur, Berlin (D)	364
West 8 Urban Design & Landscape Architecture, Rotterdam (NL)	160, 162, 174-177, 364
Wetzel & von Seht, Hamburg (D)	220-221
White Arkitekter, Stockholm (S)	160
Wiel Arets Architects, Amsterdam (NL)	364
Wiencke Architekten, Dresden (D)	294
Willmotte & Associés, Paris (F)	350
WindStone International, Berlin (D)	65
Winfried Brenne Architekten, Berlin (D)	242
Wingårdh Arkitektkontor, Göteborg (S)	138
Wörner Traxler Richter Planungsgesellschaft, Frankfurt am Main (D)	206
wulf architekten, Stuttgart (D)	11, 196, 198-199, 206, 231
wwa - wöhr heugenhauser architekten, Vienna (A)	152

X

Xaveer De Geyter Architects, Brussels (BE)	11

Y

Yassir Khalil Studio, Casablanca (MA)	68, 70-75
YI ARCHITECTS, Cologne (D)	258, 262
YO2 Architects, Seoul (ROK)	114-115
Yves WOZNIAK Architecte, Marquillies (F)	66

Z

Zaha Hadid Architects, London (UK)	10, 11
zanderroth architekten, Berlin (D)	242, 248, 250-251
Zeytinoglu Architects, Vienna (A)	364, 366
Zine El Abidine Lasouini, Fez (MA)	68
Zinterl Architekten, Graz (A)	186, 190
ziya necati özkan architectural & engineering office, Nikosia (CY)	66
Zvi Hecker Architekt, Berlin (D)	294

方案设计者及合作者均来自 [phase eins]. 工作室：

Christine Eichelmann, Benjamin Hossbach, Christian Lehmhaus

参与者：

Uwe Barsch, Alexander Bulgrin, Christina D.hmen, Julia Feier, Barbara Frei, Teresa Go, Stefan Haase, Raschid Hafiz, Marc Havekost, Svea Heinemann, Sylvia Hofmann, Sebastian Illig, Maja Kastaun, Ronny Kutter, Susanne Mocka, Brigitte Panek, Angela Salzburg, Stefan Sa., Daniel Schoene, Bj.rn Steinhagen, Harald Theiss, Ariane Vetter, Bettine Volk, Gernot Würtenberger, Sibel Yilmaz

自由职业者、学生和实习生：

Mogdeh Ali, Katrin Bade, Mario B.r, Lukasz Baran, Weronika Bartkowiak, Rahaf Bata, Philip Baumbach, Tim Boczek, Aljoscha Boesser, Jens Drexhage, Lana Eichelmann, Luis Eichelmann, Nicole Erbe, Karina Ernst, Niklas Fissel, Serin Geambazu, Melanie Glasenapp, Dorothee Haas, Felix Hiller, Paula Hentschel, Fatemeh Irandoost, Michael Kandel, Levan Kikna, Marika Kunert, Patrick Kutterolf, Judith Lehmhaus, Vera Lehmhaus, Josefine Mochar, Sandja M.ckel, Saskia Müller, Olrik Neubert, Anne Peters, Julien Schwindenhammer, Julia Trapp, Leonie Treseler, Jakob Ulbrych, Yakub Vardar, Tra Mi Vu Pham, Sandra Warken, Kai Woog, Anyana Zimmermann

初赛评审者：

Annette Bresinsky, Heinrich Burchard (.), Tobias Buschbeck, Georg Düx, Friedhelm Gülink, Helmut Hanle, Konstanze Herrmann, David Meyer, Birgit Petersen, Alexander G. Williams, Birgit Wolf

合作公司：

Michael R.dler (exhibition construction and display stands), Berlin
www.mraedler.de
Klaus Rupprecht (translations), Berlin
Olaf Schreiber, Olaf Schreiber Software (internet- and database solutions), Berlin
www.schreiber.biz
Sirko Sparing, Fa. dBusiness (copy and printing), Berlin
www.dbusiness.de
Hans-Joachim Wuthenow, (photography), Berlin
www.wuthenow-foto.de

致谢：

我们向各个参赛作品的业主表示衷心的感谢，他们在参赛作品材料的收集与编写方面提供了莫大的支持与帮助；我们向 [phase eins]. 工作室的成员表示诚挚的感谢，他们的辛勤工作使各个参赛作品的成果展示变得富有逻辑、生动有趣且创意十足；我们向参与本书策划与编写的其他人员表示感谢，感谢他们为我们提供了努力的方向与前进的动力。